OPTICAL METHODS OF MEASUREMENT

OPTICAL ENGINEERING

Series Editor

Brian J. Thompson

Distinguished University Professor
Professor of Optics
Provost Emeritus
University of Rochester
Rochester, New York

1. Electron and Ion Microscopy and Microanalysis: Principles and Applications, *Lawrence E. Murr*
2. Acousto-Optic Signal Processing: Theory and Implementation, *edited by Norman J. Berg and John N. Lee*
3. Electro-Optic and Acousto-Optic Scanning and Deflection, *Milton Gottlieb, Clive L. M. Ireland, and John Martin Ley*
4. Single-Mode Fiber Optics: Principles and Applications, *Luc B. Jeunhomme*
5. Pulse Code Formats for Fiber Optical Data Communication: Basic Principles and Applications, *David J. Morris*
6. Optical Materials: An Introduction to Selection and Application, *Solomon Musikant*
7. Infrared Methods for Gaseous Measurements: Theory and Practice, *edited by Joda Wormhoudt*
8. Laser Beam Scanning: Opto-Mechanical Devices, Systems, and Data Storage Optics, *edited by Gerald F. Marshall*
9. Opto-Mechanical Systems Design, *Paul R. Yoder, Jr.*
10. Optical Fiber Splices and Connectors: Theory and Methods, *Calvin M. Miller with Stephen C. Mettler and Ian A. White*
11. Laser Spectroscopy and Its Applications, *edited by Leon J. Radziemski, Richard W. Solarz, and Jeffrey A. Paisner*
12. Infrared Optoelectronics: Devices and Applications, *William Nunley and J. Scott Bechtel*
13. Integrated Optical Circuits and Components: Design and Applications, *edited by Lynn D. Hutcheson*
14. Handbook of Molecular Lasers, *edited by Peter K. Cheo*
15. Handbook of Optical Fibers and Cables, *Hiroshi Murata*
16. Acousto-Optics, *Adrian Korpel*

17. Procedures in Applied Optics, *John Strong*
18. Handbook of Solid-State Lasers, *edited by Peter K. Cheo*
19. Optical Computing: Digital and Symbolic, *edited by Raymond Arrathoon*
20. Laser Applications in Physical Chemistry, *edited by D. K. Evans*
21. Laser-Induced Plasmas and Applications, *edited by Leon J. Radziemski and David A. Cremers*
22. Infrared Technology Fundamentals, *Irving J. Spiro and Monroe Schlessinger*
23. Single-Mode Fiber Optics: Principles and Applications, Second Edition, Revised and Expanded, *Luc B. Jeunhomme*
24. Image Analysis Applications, *edited by Rangachar Kasturi and Mohan M. Trivedi*
25. Photoconductivity: Art, Science, and Technology, *N. V. Joshi*
26. Principles of Optical Circuit Engineering, *Mark A. Mentzer*
27. Lens Design, *Milton Laikin*
28. Optical Components, Systems, and Measurement Techniques, *Rajpal S. Sirohi and M. P. Kothiyal*
29. Electron and Ion Microscopy and Microanalysis: Principles and Applications, Second Edition, Revised and Expanded, *Lawrence E. Murr*
30. Handbook of Infrared Optical Materials, *edited by Paul Klocek*
31. Optical Scanning, *edited by Gerald F. Marshall*
32. Polymers for Lightwave and Integrated Optics: Technology and Applications, *edited by Lawrence A. Hornak*
33. Electro-Optical Displays, *edited by Mohammad A. Karim*
34. Mathematical Morphology in Image Processing, *edited by Edward R. Dougherty*
35. Opto-Mechanical Systems Design: Second Edition, Revised and Expanded, *Paul R. Yoder, Jr.*
36. Polarized Light: Fundamentals and Applications, *Edward Collett*
37. Rare Earth Doped Fiber Lasers and Amplifiers, *edited by Michel J. F. Digonnet*
38. Speckle Metrology, *edited by Rajpal S. Sirohi*
39. Organic Photoreceptors for Imaging Systems, *Paul M. Borsenberger and David S. Weiss*
40. Photonic Switching and Interconnects, *edited by Abdellatif Marrakchi*
41. Design and Fabrication of Acousto-Optic Devices, *edited by Akis P. Goutzoulis and Dennis R. Pape*
42. Digital Image Processing Methods, *edited by Edward R. Dougherty*
43. Visual Science and Engineering: Models and Applications, *edited by D. H. Kelly*
44. Handbook of Lens Design, *Daniel Malacara and Zacarias Malacara*
45. Photonic Devices and Systems, *edited by Robert G. Hunsperger*
46. Infrared Technology Fundamentals: Second Edition, Revised and Expanded, *edited by Monroe Schlessinger*
47. Spatial Light Modulator Technology: Materials, Devices, and Applications, *edited by Uzi Efron*
48. Lens Design: Second Edition, Revised and Expanded, *Milton Laikin*
49. Thin Films for Optical Systems, *edited by François R. Flory*

50. Tunable Laser Applications, *edited by F. J. Duarte*
51. Acousto-Optic Signal Processing: Theory and Implementation, Second Edition, *edited by Norman J. Berg and John M. Pellegrino*
52. Handbook of Nonlinear Optics, *Richard L. Sutherland*
53. Handbook of Optical Fibers and Cables: Second Edition, *Hiroshi Murata*
54. Optical Storage and Retrieval: Memory, Neural Networks, and Fractals, *edited by Francis T. S. Yu and Suganda Jutamulia*
55. Devices for Optoelectronics, *Wallace B. Leigh*
56. Practical Design and Production of Optical Thin Films, *Ronald R. Willey*
57. Acousto-Optics: Second Edition, *Adrian Korpel*
58. Diffraction Gratings and Applications, *Erwin G. Loewen and Evgeny Popov*
59. Organic Photoreceptors for Xerography, *Paul M. Borsenberger and David S. Weiss*
60. Characterization Techniques and Tabulations for Organic Nonlinear Optical Materials, *edited by Mark Kuzyk and Carl Dirk*
61. Interferogram Analysis for Optical Testing, *Daniel Malacara, Manuel Servín, and Zacarias Malacara*
62. Computational Modeling of Vision: The Role of Combination, *William R. Uttal, Ramakrishna Kakarala, Sriram Dayanand, Thomas Shepherd, Jagadeesh Kalki, Charles F. Lunskis, Jr., and Ning Liu*
63. Microoptics Technology: Fabrication and Applications of Lens Arrays and Devices, *Nicholas F. Borrelli*
64. Visual Information Representation, Communication, and Image Processing, *Chang Wen Chen and Ya-Qin Zhang*
65. Optical Methods of Measurement: Wholefield Techniques, *Rajpal S. Sirohi and Fook Siong Chau*

Additional Volumes in Preparation

Integrated Optical Circuits and Components: Design and Applications, *edited by Edmund J. Murphy*

Computational Methods for Electromagnetic and Optical Systems, *John M. Jarem and Partha P. Banerjee*

Adaptive Optics Engineering Handbook, *edited by Robert K. Tyson*

OPTICAL METHODS OF MEASUREMENT

WHOLEFIELD TECHNIQUES

RAJPAL S. SIROHI

Indian Institute of Technology
Madras, India

FOOK SIONG CHAU

The National University of Singapore
Singapore

MARCEL DEKKER, INC. NEW YORK • BASEL

ISBN: 0-8247-6003-4

This book is printed on acid-free paper.

Headquarters
Marcel Dekker Inc.
270 Madison Avenue, New York, NY 10016
tel: 212-696-9000; fax: 212-685-4540

Eastern Hemisphere Distribution
Marcel Dekker AG
Hutgasse 4, Postfach 812, CH-4001 Basel, Switzerland
tel: 41-61-261-8482; fax: 41-61-261-8896

World Wide Web
http://www/dekker.com

The publisher offrers discounts on this book when ordered in bulk quantities. For more information, write to Special Sales/Professional Marketing at the headquarters address above.

Current printing (last digit):
10 9 8 8 7 6 5 4 3 2 1

PRINTED IN THE UNITED STATES OF AMERICA

To
Those
Who Make Things Happen

From the Series Editor

Over the last 20 years, optical science and engineering has emerged as a discipline in its own right. This occurred with the realization that we are dealing with an integrated body of knowledge that has resulted in optical science, engineering and technology becoming an enabling discipline that can be applied to a variety of scientific, industrial, commercial, and military problems to produce operating devices and hardware systems. This series of books is a testament to the truth of the preceding statement.

Quality control and the testing that supports that control have become essential tools in modern industry and modern laboratory processes. Optical methods of measurement have provided many of the essential techniques of current practice. This current volume on Optical Methods of Measurement emphasizes wholefield measurement methods as opposed to point measurement, i.e., sensing a field all at once and then mapping that field for the parameter or parameters under consideration as contrasted to building up that field information by a time series of point-by-point determinations. The initial output of these wholefield systems of measurement is often a fringe pattern that is then processed. The required fringes are formed by direct interferometry, holographic interferometry, phase-shifting methods, heterodyning, speckle pattern interferometry, and moiré techniques.

The methods described are applicable to many measurement scenarios, although the examples focus on modern experimental mechanics. Since this volume covers the variety of techniques available and their range of applicability together with their sensitivity and accuracy as determined by the underlying principle, the reader will find these pages an excellent guide practice of wholefield measurement as well as a desk reference volume.

Brian J. Thompson

Preface

Optical techniques of measurement are among the most sensitive known today. In addition, they are noncontact, noninvasive, and fast. In recent years, the use of optical techniques for measurement has dramatically increased, and applications range from determining the topography of landscapes to checking the roughness of polished surfaces.

Any of the characteristics of a light wave — amplitude, phase, length, frequency, polarization, and direction of propagation — can be modulated by the measurand. On demodulation, the value of the measurand at a spatial point and at a particular time instant can be obtained. Optical methods can effect measurement at discrete points or over the whole field with extremely fine spatial resolution. For many applications, wholefield measurements are preferred.

This book contains a wealth of information on wholefield measurement methods, particularly those employed frequently on modern experimental mechanics since the variable that is normally monitored is displacement. Thus, the methods can be used to determine surface deformation, strains, and stresses. By extension, they can be applied in the nondestructive evaluation of components. There is no doubt that the methods described are applicable to other fields as well such as the determination of surface contours and surface roughness. With the appropriate setup, these wholefield optical methods can be used to obtain material properties and temperature and pressure gradients. Throughout the book, emphasis is placed on the physics of the techniques, with demonstrative examples of how they can be applied to tackle particular measurement problems.

Any optical technique has to involve a light source, beam-handling optics, a detector, and a data-handling system. In many of the wholefield measurement techniques, a laser is the source of light. Since much information on lasers is available, this aspect is not discussed in the book. Instead, we have included an introductory chapter on the concept of waves and associated phenomena. The propagation of waves is discussed in Chapter 2.

Chapter 3 deals with current phase evaluation techniques, since almost all the techniques described here display the information in the form of a fringe pattern. This fringe pattern is evaluated to a high degree of accuracy using currently available methods to extract the information of interest. The various detectors available for recording wholefield information are described in Chapter 4. Thus the first four chapters provide the background for the rest of the book.

Chapter 5 is on holographic interferometry. A number of techniques have been included to give readers a good idea of how various applications may be handled. The speckle phenomenon, although a bane of holographers, has emerged as a very good measurement technique. Several techniques are now available to measure the in-plane and out-of-plane displacement components, tilts, or slopes. In Chapter 6, these techniques are contrasted with those based on holographic interferometry, with discussion of their relative advantages and areas of applicability. In recent times, electronic detection in conjunction with phase evaluation methods have been increasingly used. This makes all these speckle techniques almost real-time approaches, and they have consequently attracted more attention from industry. Chapter 6 includes descriptions of various measurement techniques in speckle metrology.

Photoelasticity is another wholefield measurement technique that has existed for some time. Experiments are conducted on transparent models that become birefringent when subjected to load. Unlike holographic interferometry and speckle-based methods, which measure displacement or strain, photoelasticity gives the stress distribution directly. Photoelasticity, including the technique of holophotoelasticity, is covered in Chapter 7.

The moiré technique, along with the Talbot phenomenon, is a powerful measurement technique and covers a very wide range of sensitivity. Chapter 8 presents various techniques, spanning the measurement range from geometrical moiré with low-frequency gratings to the high-sensitivity moiré with very high frequency diffraction gratings.

The book has evolved as a result of the authors' involvement with research and teaching of these techniques over more than two decades. It puts together the major wholefield methods of measurement in one text and therefore should be useful to students taking a graduate course in optical experimental mechanics and to industrialists who are interested in investigating the techniques available and the physics behind them. It will also serve as a useful reference for researchers in the field.

We acknowledge the strong support given by Professor Brian J. Thompson and the excellent work of the staff of Marcel Dekker, Inc. This book would not have been possible without the constant support and encouragement of our wives, Vijayalaxmi Sirohi and Chau Swee Har, and our families; to them, we express our deepest thanks.

Rajpal S. Sirohi
Fook Siong Chau

Contents

OPTICAL METHODS OF MEASUREMENT

Chapter 1

Waves

Optics pertains to the generation, amplification, propagation, reception, modification, and modulation of light, and light is considered to be a very tiny portion of the electromagnetic spectrum responsible for vision. In short, optics is science, technology, and engineering with light. Measurement techniques based on light fall under the heading of optical metrology. Therefore, optical metrology comprises varieties of measurement techniques, which include dimensional measurements, measurement of process variables, and measurement of electrical quantities like current and voltages.

Optical methods are noncontact and noninvasive. Indeed the light does interact with the measurand but in the majority of cases does not influence its value; hence the measurements are termed noncontact. Furthermore, information about the whole object can be obtained simultaneously, thereby giving a capability of whole field measurement. The measurements are not influenced by electromagnetic fields. Light wavelength, moreover, is a measurement stick in length metrology. Some of these attributes make the optical methods indispensable for measurements of several different kinds.

I. THE WAVE EQUATION

Light is an electromagnetic wave and therefore should satisfy the wave equation

$$\nabla^2\mathbf{E}(r, t) - \frac{1}{c^2}\frac{\partial^2\mathbf{E}(r, t)}{\partial t^2} = 0 \tag{1}$$

where $\mathbf{E}(r, t)$ is the instantaneous light field and c is the velocity of light. This is a scalar wave equation. We are interested in the solutions of the wave equation [Eq. (1)] that represent monochromatic waves, that is,

$$\mathbf{E}(r, t) = \mathbf{E}(r)e^{-i\omega t} \tag{2}$$

where $\mathbf{E}(r)$ is the amplitude and ω is the circular frequency of the wave. Substituting Eq. (2) into Eq. (1) leads to

$$\nabla^2\mathbf{E}(r) + k^2\mathbf{E}(r) = 0 \tag{3}$$

where $k^2 = \omega^2/c^2$. This is the well-known Helmholtz equation. A solution of the Helmholtz equation (Eq. 3) for $\mathbf{E}(r)$ will provide a monochromatic solution of the wave equation.

II. THE PLANE WAVE

There are several solutions of the wave equation; one of the solutions is of the form

$$\mathbf{E}(r) = \mathbf{E}_0 \exp(i\mathbf{k} \cdot \mathbf{r}) \tag{4}$$

This represents a plane wave of infinite cross section and having a constant amplitude $\mathbf{E}_0$. A plane monochromatic wave spans the whole space for all times and is only a mathematical realization. A real plane wave, called a collimated wave, is limited in its transverse dimensions. The limitation may be imposed by optical elements or systems, or by physical stops. This restriction on transverse dimensions of the wave leads to the phenomenon of diffraction, a topic described in more detail in Chapter 2.

III. THE SPHERICAL WAVE

Another solution of Eq. (3) is

$$E(r) = \frac{A}{r}\exp(i\,\mathrm{k}\cdot\mathrm{r}) \tag{5}$$

where A is a constant that represents the amplitude of the wave at unit distance. This solution represents a spherical wave. A plane wave propagates in a particular direction, while a spherical wave is not unidirectional. A quadratic approximation to the spherical wave is

$$E\,(x, y, z) = \frac{A}{z}e^{ikz}\exp\left[\frac{ik}{2z}(x^2 + y^2)\right] \tag{6}$$

This represents the amplitude at any point (x, y, z) on a plane distant z from the point source. This expression is very often used in discussions of propagation through an optical system. A collimated wave is obtained by placing a point source (source size diffraction limited) at the focal point of a lens. This is in fact a truncated plane wave. A spherical wave emanates from a point source or it may converge to a point.

IV. THE CYLINDRICAL WAVE

Often, a fine slit is illuminated by a broad source. The waves emanating from such sources are called cylindrical waves. The surfaces of constant phase, far away from the source, are cylindrical. A cylindrical lens placed in a collimated beam will generate the cylindrical wave that would converge to a line focus. The amplitude distribution for cylindrical waves, far away from the narrow slit, can be written as follows:

$$E(r) = \frac{A}{\sqrt{r}} exp(ikr) \tag{7}$$

Figure 1 shows these waves schematically.

V. THE LASER BEAM

A laser beam is a pointed beam that propagates as a nearly unidirectional wave with little divergence. The beam cross section is

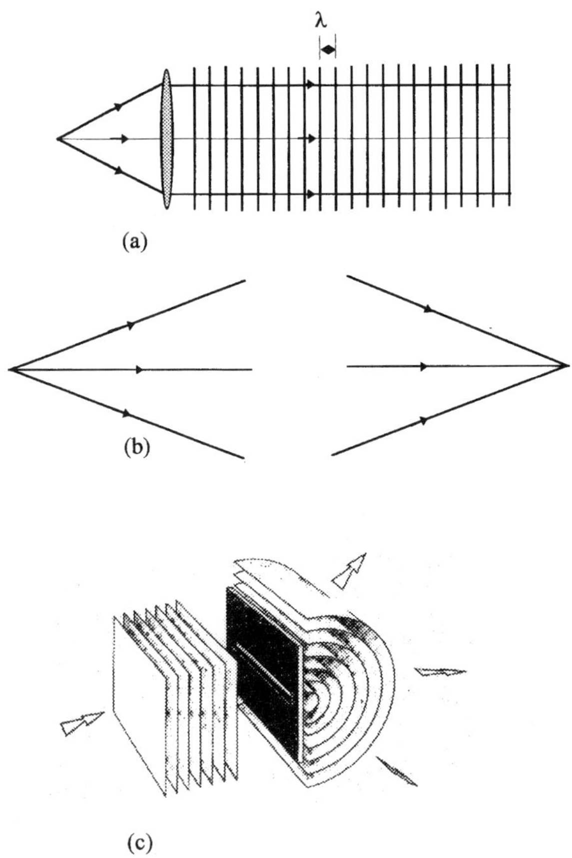

Figure 1 (a) Plane wave. (b) Diverging and converging spherical waves. (c) Cylindrical wave. (Fig. c by permission from *Optics* by E. Hecht and A. Zajac, Addison-Wesley, 1974.)

finite. Let us now seek the solution of Eq. (3) the Helmholtz equation, which represents a beam. We therefore write a solution in the form

$$E(r) = E_0(r)e^{ikz} \tag{8}$$

This solution differs from the plane wave propagating along the z direction in that its amplitude $E_0(r)$ is not constant. Furthermore, the solution should represent a beam; that is, it should possess unidirectionality and should have a finite cross section. The variations of $E_0(r)$ and $\partial E_0(r)/\partial z$ over a distance of the order of the wavelength along the z direction are assumed to be negligible. This implies that the field varies approximately as e^{ikz} over a distance of a few wavelengths. When the field $E_0(r)$ meets these approximations, the Helmholtz equation takes the form

$$\nabla_T^2 E_0 + 2ik\frac{\partial E_0}{\partial z} = 0 \tag{9}$$

where ∇_T^2 is the transverse Laplacian. Equation (9), known as the paraxial wave equation, is a consequence of weak factorization represented by Eq. (8) and other assumptions mentioned earlier.

VI. THE GAUSSIAN BEAM

The intensity distribution in a beam from a laser oscillating in TEM_{00} mode is given by

$$I(r) = I_0 \exp\left(\frac{-2r^2}{w^2}\right) \tag{10}$$

where $r = (x^2 + y^2)^{1/2}$. The intensity $I(r)$ drops to $1/e^2$ value of the peak intensity I_0 at a distance $r = w$. The parameter w, called the spot size of the beam, depends on the z coordinate. The intensity profile is Gaussian; hence the beam is called a Gaussian beam. The Gaussian beam has a very special place in optics and is often used for measurement. The beam maintains its profile as it propagates.

The solution of Eq. (9), which represents the Gaussian beam, can be written as follows:

$$
\begin{aligned}
E(r) &= A\frac{\exp[-r^2/w^2(z)]}{[1+(\lambda z/\pi w_0^2)]^{1/2}}\exp\left[\frac{ikr^2}{2R(z)}\right] \\
&\quad \exp\left[-i\tan^{-1}\left(\frac{\lambda z}{\pi w_0^2}\right)\right]\exp(ikz) \\
&= A\frac{w_0}{w(z)}\exp\left[-\frac{r^2}{w^2(z)}\right]\exp\left[\frac{ikr^2}{2R(z)}\right] \\
&\quad \exp\left\{i\left[kz-\tan^{-1}\left(\frac{z}{z_0}\right)\right]\right\}
\end{aligned} \tag{11}
$$

where A, w_0, and z_0 are constants. The constant z_0 is called the Rayleigh range. It is related to the spot size w_0 at $z = 0$ plane as $z_0 = \pi w_0^2/\lambda$. The radius of curvature $R(z)$ and spot size $w(z)$ of the beam at an arbitrary z plane are given as follows:

$$
R(z) = z\left[1+\left(\frac{\pi w_0^2}{\lambda z}\right)^2\right] = z\left[1+\left(\frac{z_0}{z}\right)^2\right] \tag{12a}
$$

and

$$
w(z) = w_0\left[1+\left(\frac{\lambda z}{\pi w_0^2}\right)^2\right]^{1/2} = w_0\left[1+\left(\frac{z}{z_0}\right)^2\right]^{1/2} \tag{12b}
$$

The amplitude of the wave is

$$
A\frac{w_0}{w(z)}\exp\left[-\frac{r^2}{w^2(z)}\right] \tag{13}
$$

The amplitude decreases as $\exp[-r^2/w^2(z)]$ (Gaussian beam). The spot size $w(z)$ increases as the beam propagates, and consequently the amplitude on axis decreases. The Gaussian distribution flattens. The spot size $w(z)$ at very large distance is

$$w(z)\Big|_{z\to\infty} = \frac{\lambda z}{\pi w_o}. \tag{14}$$

The half-divergence angle is obtained as follows:

$$\theta = \frac{dw}{dz} = \frac{\lambda}{\pi w_0} \tag{15}$$

The second term in Eq. (11) represents the radial phase factor with $R(z)$ as the radius of curvature of the beam. At large distances [i.e., $z \gg z_0$, $R(z) \to z$], the beam appears to originate at $z = 0$. However, the wavefront is plane at the $z = 0$ plane. In general, $z =$ constant plane is not an equiphasal plane. The last term in Eq. (11) represents the propagation phase. Figure 2 shows the variation of the amplitude as the beam propagates.

The intensity distribution in the Gaussian beam is

$$I(r, z) = |A|^2 \frac{w_0^2}{w^2(z)} \exp\left[-\frac{2r^2}{w^2(z)}\right] \tag{16}$$

On the beam axis ($r = 0$), the intensity varies as

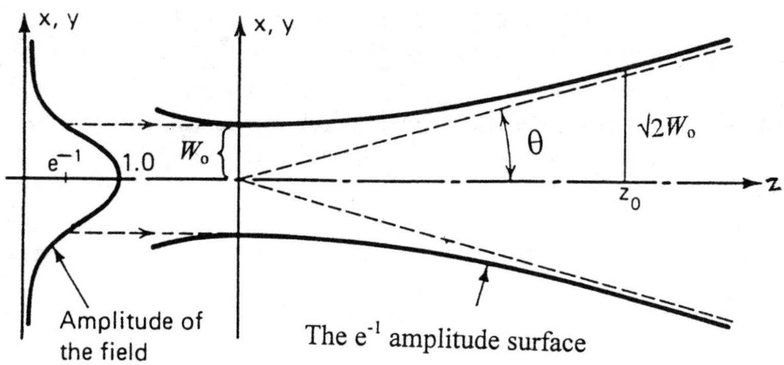

Figure 2 A Gaussian beam.

$$I(0, z) = I_0 \frac{w_0^2}{w^2(z)} = \frac{I_0}{1 + (z/z_0)^2} \tag{17}$$

It has its peak value I_0 at $z = 0$ and keeps on dropping with increasing z. The value drops to 50% of the peak value at $z = \pm z_0$.

VII. INTERFERENCE

It is sometimes said that light plus light equals darkness. The explanation for this statement lies in the phenomenon of interference. When two coherent waves are superposed, redistribution of light intensity takes place in the region of superposition. We will first define the term "intensity." It is the average value of the Poynting vector and is expressed in watts per meters squared (W/m^2). Detectors, like the eye, photographic emulsions, and photodetectors, cannot follow the amplitude variations in the light beam and respond only to energy or intensity. For this reason, the interference phenomenon becomes more important as it renders phase information into intensity information, to which the detectors respond.

Let us first consider a wave represented as

$$u(r; t) = u_0 \cos(\omega t - \mathbf{k} \cdot \mathbf{r}) \tag{18}$$

This is a scalar wave of frequency ω propagating in an arbitrary direction. A detector will measure an average value of $u^2(r; t)$. However, light being a small part of the electromagnetic spectrum propagates as transverse waves. To explicitly express the dependence of fringe contrast on the polarization of the waves, we consider the interference between two polarized waves. Let the two waves be represented by

$$\mathbf{E}_1(r; t) = \mathbf{E}_{01} \cos(\omega t - \mathbf{k} \cdot \mathbf{r}_1) \tag{19a}$$

and

$$\mathbf{E}_2(r; t) = \mathbf{E}_{02} \cos(\omega t - \mathbf{k} \cdot \mathbf{r}_2) \tag{19b}$$

When these two waves are superposed, the detector will measure the intensity obtained by the average value of $\{\mathbf{E}_1 + \mathbf{E}_2\}^2$. This can be expressed as follows:

$$I(\delta) = I_1 + I_2 + 2\mathbf{E}_{01} \cdot \mathbf{E}_{02} \cos\delta \tag{20}$$

where $\delta = \bar{k}(\bar{r}_2 - \bar{r}_1)$ is the phase difference between the two waves and I_1, I_2 are the intensities of the individual waves. It has been tacitly assumed that δ does not vary with time: the interfering waves are coherent. It is thus seen that the resultant intensity $I(\delta)$ consists of two terms $(I_1 + I_2)$ and $2\mathbf{E}_{01} \cdot \mathbf{E}_{02} \cos\delta$. The first term is the dc bias, which may be constant or varying slowly over the xy plane, while the second term is the interference term, which varies rapidly with δ. Therefore, the resultant intensity in the superposed region varies between the minimum and maximum values. This variation is due to the phenomenon of interference, which redistributes the light. The regions of maximum brightness constitute the bright fringes and those of minimum brightness, the dark fringes. The light distribution on the x–y plane is called a fringe pattern or an interference pattern. It may be noted that the fringes are loci of constant phase difference δ. Consecutive bright (or dark) fringes are separated by 2π in phase difference. Following Michelson, we define the visibility of fringes as follows:

$$V = \frac{I_{\max} - I_{\min}}{I_{\max} + I_{\min}} \tag{21}$$

where $I_{\max}$ and $I_{\min}$ are the maximum and minimum values of the resultant intensity. Substituting these values of resultant intensity into the expression for the visibility [Eq. (21)], we obtain

$$V = \frac{2\mathbf{E}_{01} \cdot \mathbf{E}_{02}}{I_1 + I_2} \tag{22}$$

The visibility depends on two parameters: the polarization states of the two beams and the intensity ratio, which is not obvious in the form of Eq. (22). The visibility will be maximum if the interfering beams have the same state of polarization (i.e., if the **E** vectors in both beams are directed in the same direction). If the beams are orthogonally polarized, no fringe pattern forms. In other words, the interference term is zero. The optical setups are thus configured accordingly. For example, the **E** vector in the beam is arranged to be vertical when the setup is on a horizontal table and the beams confined to the horizontal plane. Successive reflections at the mirrors will not rotate the plane of polarization. Hence the interfering beams are in the same state of polarization. Writing $\mathbf{E}_{01} = \sqrt{I_1}$ and $\mathbf{E}_{02} = \sqrt{I_2}$, we can express the visibility V as follows:

$$V = \frac{2\sqrt{I_1 I_2}}{I_1 + I_2} \tag{23}$$

The visibility is obviously unity when the two beams are of equal intensity. Further, the fringe visibility is quite good even if one of the beams is very weak. This explains why the contrast of spurious fringes (due to dust particles, scratches, etc.) is very good. In this analysis, it has been tacitly assumed that both the beams are coherent. The beams are therefore derived from the same parent beam. However, real sources emit partially coherent light; hence the degree of coherence also enters into the definition of the visibility.

The interfering beams can be derived from the parent beam either by wave front division or by amplitude division. One of the well-known examples of wave front division is the famous double-slit experiment by Young, as shown in Figure 3. A pair of slits samples a wave front at two spatial locations. These

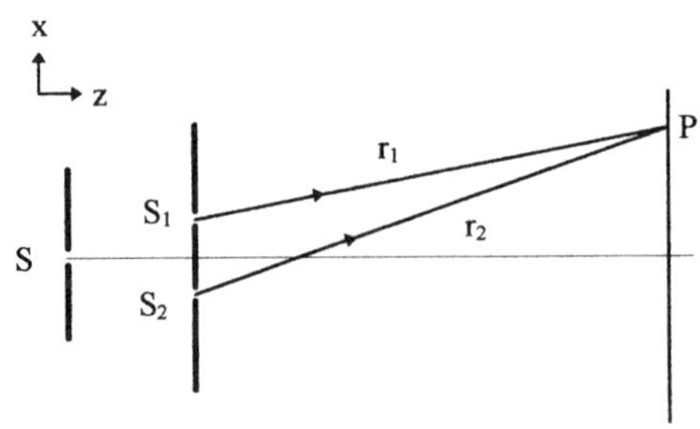

Figure 3 Young's double-slit experiment.

sampled wave fronts are diffracted by the slits, thereby expanding their spatial extent so that they superpose at a plane some distance away from the slit plane. Alternately, the pair of slits could be considered a pair of secondary sources excited by the parent wave. The phase difference between the two waves at any point on the x–y plane distant z from the slit plane, under first-order approximation, is

$$\delta = k\frac{xd}{z} \tag{24}$$

where $k = 2\pi/\lambda$ and d is the separation between the slits. The intensity distribution in the interference pattern, however, is given by

$$I(\delta) = 2I_0 \text{sinc}^2\left(\frac{kxb}{z}\right)[1 + \cos\delta] \tag{25}$$

where I_0 is the intensity on axis due to each slit and $\text{sinc}(x) = \sin(x/x)$ is the amplitude distribution due to diffraction at the slit of width $2b$. The intensity distribution is modulated by the diffraction term: if the slit is very narrow, the diffracted intensity is distributed over a very wide angular range. Here it may be remarked that gratings create multiple beams by wave front division.

One of the earliest instruments based on amplitude division is the Michelson interferometer shown in Figure 4. The path matching between the two arms of the interferometer is accomplished by introducing a compensator so that low coherence length sources could also be used. A plane parallel plate is another example of wave front division but requires moderate coherence length of the source. All polarization interferometers utilize amplitude division. The Fabry–Perot interferometer and thin film structures utilize multiple interference by amplitude division.

Let us first consider interference between two plane waves. One plane wave is incident normally at the x–y plane while the other plane wave lies in the x–z plane and makes an angle θ with the z axis. The phase difference δ is given by

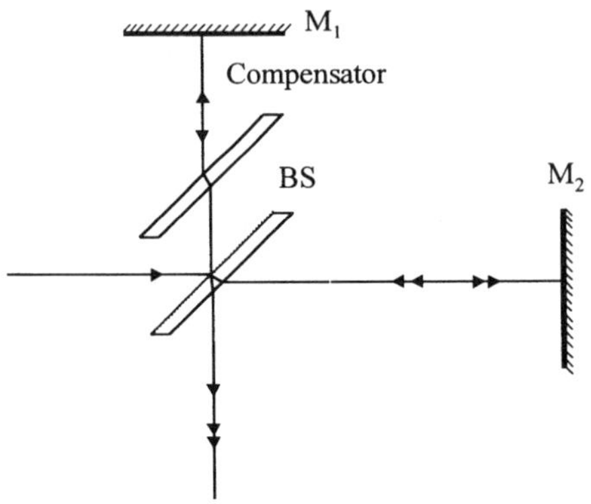

Figure 4 Schematic representation of a Michelson interferometer: BS, beam splitter; M_1, M_2, mirrors.

$$\delta = kx \sin\theta \tag{26}$$

The fringes run parallel to the y axis and the fringe spacing is $\bar{x} = \lambda/\sin\theta$. The fringe pattern is a system of straight lines running parallel to the y axis and with a constant spacing. If the plane waves were incident symmetrically about the z axis and enclosed an angle θ between them, the fringe spacing would be $\bar{x} = \lambda/2\sin(\theta/2)$. For small θ, the fringe width is almost equal in both the cases.

Let us now consider the interference between a plane wave along the z axis and a spherical wave emanating from a point source on the origin. The phase difference is given by

$$\delta = \delta_0 - k\frac{x^2 + y^2}{2z} \tag{27}$$

where δ_0 is a constant phase difference and the distance between the observation plane and the point source is z. This expression is valid under paraxial approximation.

Bright fringes are formed, assuming $\delta_0 = 0$, when

$$x^2 + y^2 = 2m\lambda z \qquad \text{for } m = 0, 1, 2, 3, \ldots \tag{28}$$

The fringe pattern consists of circular fringes with radii proportional to the square root of the natural number, and the brightness of the fringe on axis depends on δ_0 when $\delta_0 \neq 0$. Such an interference record is known as Gabor zone plate. If the plane wave is added at an angle, the fringes will appear as arcs of circles. The interference pattern between a cylindrical wave and an on-axis collimated wave is a one-dimensional zone plate. The pattern consists of straight fringes parallel to the axis of the cylinder but with decreasing spacing. The schematics for observing interference between plane waves and also between plane and spherical waves are shown in Figure 5.

VIII. INTERFERENCE BETWEEN TWO SPHERICAL WAVES

It is very illustrative to study the fringe formation due to interference between two spherical waves emanating from two point sources S_1 and S_2. It is assumed that the point sources are coherent. The phase difference between the two waves at any point P is given by

$$\delta = k(r_2 - r_1) \tag{29}$$

where r_2 and r_1 are the distances between S_1 and P, and S_2 and P, respectively. The surfaces of constant phase are hyperboloids. We are, however, interested in surfaces where the phase difference is an integral multiple of 2π, representing bright fringes. Figure 6 shows these surfaces as sections in a plane containing both the sources. The fringes are observed on an observation plane that intersects these surfaces. When the observation plane lies at location 1, the fringes are nearly equidistant. These are the famous Young's fringes observed in two beam interference. However, at location 2, the intersection of the observation plane with hyperboloid surfaces gives circular fringes with radii increasing as the square root of natural numbers. This region is known as Gabor region. At any other location 3, the fringes are curved. When the observation plane is placed at location 4, however, the observation plane lies parallel to the fringe surface, and it is more difficult to observe the

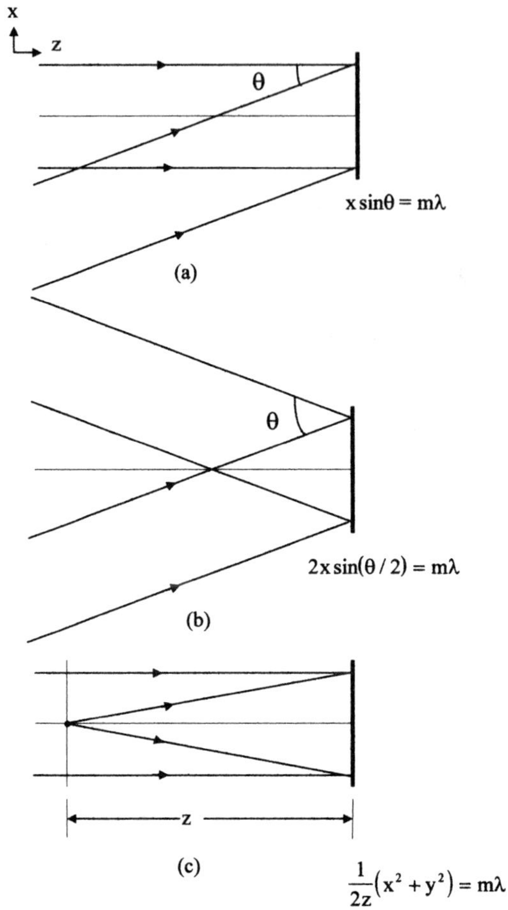

Figure 5 Interference between waves: (a) two collimated waves, (b) two symmetrical collimated waves, and (c) a collimated wave and a spherical wave.

fringes. To record these surfaces, a volume recording medium is used. This region is known as Denisyuk region. The fringe surfaces are separated by $\lambda/2$. The fringe shape and the local fringe spacing are obtained easily by substituting expressions for r_2 and r_1 in appropriate coordinate system in Eq. (29).

It was mentioned earlier that two coherent sources or waves are required for observing the interference phenomenon. These two

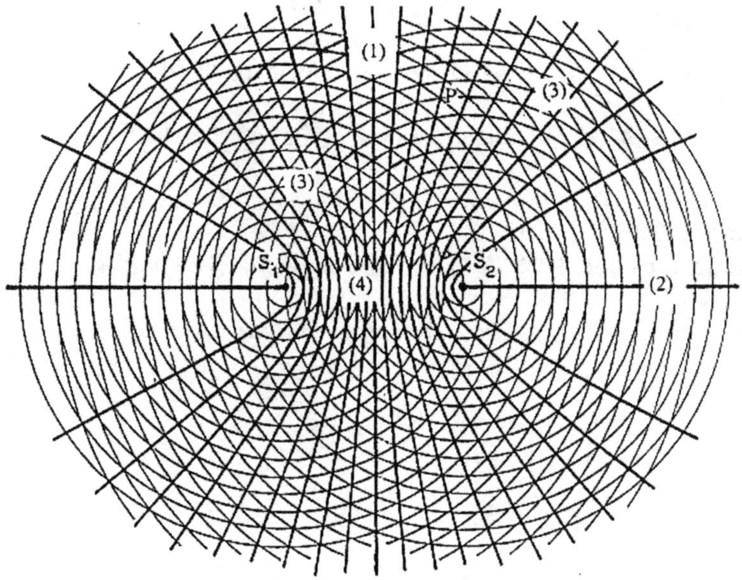

Figure 6 Curves of equal path difference of two interfering spherical waves.

waves are realized from the parent wave by amplitude division and by wave front division. Most of the interferometers (Michelson, Mach–Zehnder, etc.) utilize amplitude division. It may also be noted that the interference phenomenon is not limited to the superposition of two waves only. In fact, a large number of waves can participate in interference. The participation of more than two waves in interference is known as multiple beam interference. The participation of multiple waves modifies the intensity distribution in the fringe pattern, but the locations of the maxima of the fringes remain unchanged. Generation of multiple waves is also accomplished by both amplitude division and wavefront division. The Fabry–Perot interferometer is an example of multiple wave interference by means of amplitude division. On the other hand, gratings generate multiple waves by means of wave front division.

BIBLIOGRAPHY

1. W. T. Welford, *Geometrical Optics*, North-Holland, Amsterdam (1962).
2. W. J. Smith, *Modern Optical Engineering*, McGraw-Hill, New York (1966).
3. R. S. Longhurst, *Geometrical and Physical Optics*, Longmans, London (1967).
4. F. A. Jenkins and H. E. White, *Fundamentals of Optics*, McGraw-Hill, New York (1976).
5. R. S. Sirohi, *Wave Optics and Applications*, Orient Longmans, Hyderabad (1993).

Chapter 2

Diffraction

In an isotropic medium, a monochromatic wave propagates with its characteristic speed. For plane and spherical waves, if the phase front is known at time $t = t_1$, its location at a later time is obtained easily by multiplying the elapsed time by the velocity in the medium. The wave remains either plane or spherical. Mathematically, we can find the amplitude at any point by solving the Helmholtz equation [Eq. (3) of Chapter 1]. However, when the wave front is restricted in its lateral dimension, light is diffracted. Diffraction problems are rather complex, analytical solutions exist for only a few. The Kirchhoff theory of diffraction, though infected with self-inconsistency in the boundary conditions, yields predictions that are in close agreement with experiments.

The term "diffraction," conveniently described by Sommerfeld as "any deviation of light rays from rectilinear paths which cannot be interpreted as reflection or refraction," is often stated in a different form as well, namely: "bending of rays near the corners." Grimaldi first observed the presence of bands in the geometrical shadow region of an object. A satisfactory explanation of this observation was given by Huygens, who introduced the concept of secondary sources on the wave front at any instant and obtained the subsequent wave front as an envelope of the secondary waves. Fresnel improved on the ideas of Huygens, and the resulting theory is known as Huygens–Fresnel theory of diffraction.

This theory was placed on firmer mathematical grounds by Kirchhoff. Kirchhoff developed his mathematical theory using two assumptions about the boundary values of the light incident on the surface of the obstacle placed in the propagation path of light. This theory is known as Fresnel–Kirchhoff diffraction theory. It was found that the two assumptions of Fresnel–Kirchhoff theory are self-inconsistent. The theory, however, yields results that are in surprisingly good agreement with experiments under most circumstances. Sommerfeld modified Kirchhoff theory by eliminating one of the assumptions, and the theory is known as Rayleigh–Sommerfeld diffraction theory. Needless to say, there have been subsequent investigations by several researchers on the refinements of the diffraction theories.

I. FRESNEL DIFFRACTION

Let $u(x_0, y_0)$ be the field distribution at the aperture which lies on the x_0–y_0 plane located at $z = 0$. Under small-angle diffraction, the field at any point $P(x, y)$ (assumed to be near the optical axis) on a plane distant z from the obstacle plane can be expressed using

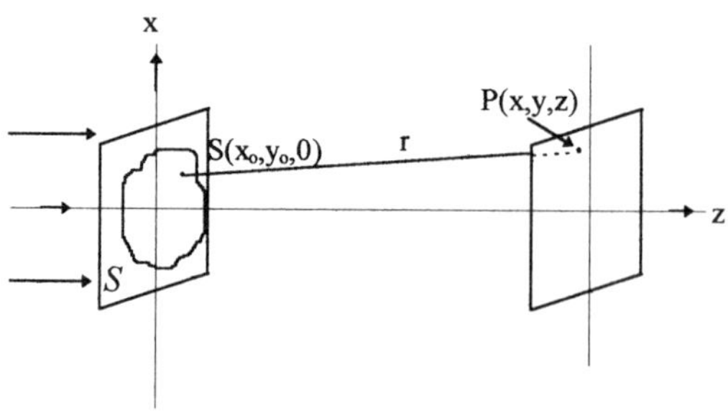

Figure 1 Diffraction at an aperture.

Fresnel–Kirchhoff diffraction theory (Fig. 1) as follows:

$$u(x, y) = \frac{1}{i\lambda z} \iint_S u(x_0, y_0) e^{ikr} \, dx_0 dy_0 \tag{1}$$

where the integral is over the area S of the aperture, and $k = 2\pi/\lambda$ is the propagation vector of the light. We can expand r in the phase term in Taylor's series as follows:

$$r = z + \frac{(x - x_0)^2 + (y - y_0)^2}{2z} - \frac{[(x - x_0)^2 + (y - y_0)^2]^2}{8z^3} + \cdots. \tag{2}$$

If we impose a condition that

$$\frac{[(x - x_0)^2 + (y - y_0)^2]^2}{8z^3} \ll \lambda \tag{3}$$

we are in the Fresnel diffraction region. The Fresnel region may extend from very near to the aperture to infinity. Condition (3) implies that the path variation at point $P(x, y)$ as point $S(x_0, y_0)$ spans the whole area S is much less than one wavelength. The amplitude $u(x, y)$ in the Fresnel region is given by

$$u(x, y) = \frac{e^{ikz}}{i\lambda z} \iint_S u(x_0, y_0) \exp\left\{\frac{ik}{2z}\left[(x - x_0)^2 + (y - y_0)^2\right]\right\} dx_0 dy_0 \tag{4}$$

Since we are invariably in the Fresnel region in most of the situations that arise in optics, Eq. (4) is frequently used.

II. FRAUNHOFER DIFFRACTION

A more stringent constraint may be placed on the distance z and/or on the size of the obstacle or aperture by setting

$$\frac{x_0^2 + y_0^2}{2z} \ll \lambda \tag{5}$$

This is the far-field condition, and when this condition is met we are in the Fraunhofer region. When condition (5) is satisfied, the amplitude $u(x, y)$ is given by

$$u(x, y) = \frac{e^{ikz}}{i\lambda z} \exp\left[\frac{ik}{2z}(x^2 + y^2)\right] \iint_S u(x_0, y_0) \exp\left[-\frac{ik}{z}(xx_0 + yy_0)\right] dx_0 dy_0 \tag{6}$$

Let us consider the illumination of the aperture by a plane wave of amplitude A. The transmittance function $t(x_0, y_0)$ of the aperture may be defined as follows:

$$t(x_0, y_0) = u(x_0, y_0)/A$$

Therefore, the field $u(x, y)$ may be expressed as follows:

$$u(x, y) = \frac{Ae^{ikz}}{i\lambda z} \exp\left[\frac{ik}{2z}(x^2 + y^2)\right] \int_{-\infty}^{\infty} \int_{-\infty}^{\infty} t(x_0, y_0) \exp\left[-\frac{ik}{z}(xx_0 + yy_0)\right] dx_0\, dy_0 \tag{7}$$

The limits of integration have been changed to $\pm\infty$ because $t(x_0, y_0)$ has a finite value within the opening and zero outside. It is seen that $u(x, y)$ can be expressed as a Fourier transform of the transmittance function. The Fourier transform is evaluated at the spatial frequencies $\mu = (x/\lambda z)$ and $\nu = (y/\lambda z)$.

III. ACTION OF A LENS

A plane wave incident on a lens is transformed to a converging/diverging spherical wave; a converging spherical wave will come to focus. Therefore, a lens may be described by a transmittance function $\tau(x_L, x_L)$, which is represented as follows:

$$\tau(x_L, y_L) = e^{i\phi_0} \exp\left[\pm \frac{ik}{2f}(x_L^2 + y_L^2)\right] \qquad \text{for } (x_L^2 + y_L^2) \leq R^2 \quad (8)$$

where ϕ_0 is a constant phase, $(k/2f)(x_L^2 + y_L^2)$ is a quadratic approximation to a spherical wave, and $2R$ is the diameter of the lens aperture. The minus sign corresponds to a positive lens. The lens is assumed to be transparent; that is, it does not introduce any attenuation.

IV. IMAGE FORMATION AND FOURIER TRANSFORMATION BY A LENS

We consider the geometry shown in Figure 2. A transparency of amplitude transmittance $t(x_0, y_0)$ is illuminated by a diverging spherical wave emanating from a point source S_0. We wish to obtain the amplitude distribution at a plane distant d_2 from the lens. This is done by invoking Fresnel–Kirchhoff diffraction theory twice in succession—first at the transparency plane and then at the lens plane. The amplitude at any point $P(x, y)$ on the observation plane is expressed as follows:

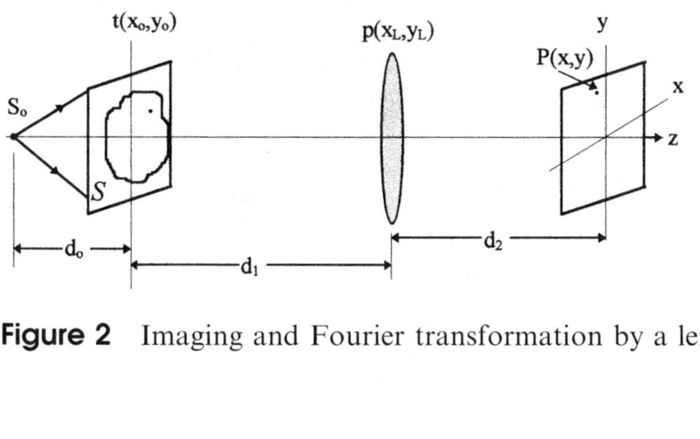

Figure 2 Imaging and Fourier transformation by a lens.

$$u(x, y) = -\frac{A\exp[ik(d_0 + d_1 + d_2)]}{\lambda^2 d_0 d_1 \mathrm{d}_2} \exp\left[\frac{ik}{2d_2}(x^2 + y^2)\right] \iint \iint p(x_\mathrm{L}, y_\mathrm{L})$$
$$\exp\left[\frac{ik}{2}(x_\mathrm{L}^2 + y_\mathrm{L}^2)\left(\frac{1}{d_1} + \frac{1}{d_2} - \frac{1}{f}\right)\right]$$
$$t(x_0, y_0) \exp\left[\frac{ik}{2}(x_0^2 + y_0^2)\left(\frac{1}{d_0} + \frac{1}{d_1}\right)\right]$$
$$\exp\left\{-ik\left[x_\mathrm{L}\left(\frac{x_0}{d_1} + \frac{x}{d_2}\right)\right.\right.$$
$$\left.\left. + y_\mathrm{L}\left(\frac{y_0}{d_1} + \frac{y}{d_2}\right)\right]\right\} dx_0 dy_0 dx_\mathrm{L} dy_\mathrm{L} \tag{9}$$

where we have introduced the lens pupil function $p(x_\mathrm{L}, y_\mathrm{L})$ as follows:

$$p(x_\mathrm{L}, y_\mathrm{L}) = 1 \quad \text{for } (x_\mathrm{L}^2 + y_\mathrm{L}^2)^{1/2} \le R$$
$$= 0 \quad \text{otherwise}$$

The limits of integration on the lens plane are therefore taken as $\pm\infty$. Writing $1/\varepsilon = 1/d_1 + 1/d_2 - 1/f$, we evaluate the integral over the lens plane in Eq. (9), assuming $P(x_\mathrm{L}, y_\mathrm{L}) = 1$ over the infinite plane. This holds valid provided the entire diffracted field from the transparency has been accepted by the lens—in other words, as long as there is no diffraction at the lens aperture. On substitution of the result thus obtained, we can obtain the amplitude $u(x, y)$ as follows:

$$u(x, y) = -\frac{A\exp[ik(d_0 + d_1 + d_2)]}{\lambda^2 d_0 d_1 \mathrm{d}_2} \varepsilon(1+i)^2 \exp\left[\frac{ik}{2d_2}(x^2 + y^2)\left(1 - \frac{\varepsilon}{d_2}\right)\right]$$
$$\iint t(x_0, y_0) \exp\left[\frac{ik}{2}(x_0^2 + y_0^2)\left(\frac{1}{d_0} + \frac{1}{d_1} - \frac{\varepsilon}{d_1^2}\right)\right]$$
$$\exp\left[-\frac{ik\varepsilon}{d_1 d_2}(xx_0 + yy_0)\right] dx_0 dy_0 \tag{10}$$

We now examine Eq. (10) to see under what conditions the x–y plane is an image plane or a Fourier transform plane of an x_0–y_0 plane.

A. Image Formation

It is known in geometrical optics that an image is formed when the imaging condition $1/d_1 + 1/d_2 - 1/f = 0$ is satisfied. We invoke this condition to see whether the amplitude distribution at the x–y plane is functionally similar to that existing at the x_0–y_0 plane. We begin with Eq. (9), invoke this condition, and evaluate the integral over the lens plane assuming no diffraction. Thus we obtain

$$\begin{aligned} u(x, y) = &-\frac{A\exp[ik(d_0 + d_1 + d_2)]}{d_0 M}\exp\left[\frac{ik}{2d_2}(x^2 + y^2)\right] \\ &\exp\left[\frac{ik}{2M^2}(x^2 + y^2)\left(\frac{1}{d_0} + \frac{1}{d_1}\right)\right] t\left(-\frac{x}{M}, -\frac{y}{M}\right) \\ = &\frac{C}{M} t\left(-\frac{x}{M}, -\frac{y}{M}\right) \end{aligned} \tag{11}$$

where $M(= d_2/d_1)$ is the magnification of the system and C is a complex constant. The amplitude $u(x, y)$ at the observation plane is identical to that of the input transparency except for the magnification. Hence, an image of the transparency is formed on the plane that satisfies the condition $1/d_1 + 1/d_2 = 1/f$.

B. Fourier Transformation

If the quadratic term in the integral in Eq. (10) is zero, the amplitude $u(x, y)$ is a two-dimensional Fourier transform of the function $t(x_0, y_0)$ multiplied by a quadratic phase factor, provided the limits of integration extend from $-\infty$ to ∞. For the quadratic term to vanish, the following condition should be satisfied:

$$\frac{1}{d_0} + \frac{1}{d_1} - \frac{\varepsilon}{d_1^2} = 0 \tag{12}$$

This condition is satisfied when $d_0 = \infty$ and $d_2 = f$. Thus, the Fourier transform of the transparency $t(x_0, y_0)$ illuminated by a collimated beam is observed at the back focal plane of the lens. It may be noted that a lens does not take the Fourier transform but merely brings it from the far field to the back focal plane.

The amplitude $u(x, y)$ when the transparency is illuminated by a collimated beam of amplitude A_p is given by

$$u(x, y) = -\frac{A_\mathrm{p}}{\lambda f} \exp[ik(d_1 + f)]\,(1 + i)^2 \exp\left[\frac{ik}{2f}(x^2 + y^2)\left(1 - \frac{d_1}{f}\right)\right] \iint t(x_0, y_0) \exp\left[-\frac{ik}{2f}(xx_0 + yy_0)\right] dx_0 dy_0 \tag{13}$$

The position-dependent quadratic phase term in front of the integral signs will also vanish if the transparency is placed at the front focal plane of the lens. The amplitude is then given by

$$u(x, y) = -u_0 \iint t(x_0, y_0) \exp\left[-\frac{ik}{2f}(xx_0 + yy_0)\right] dx_0 dy_0 \tag{14}$$

Therefore, we arrive at the following conclusions:

1. The Fourier transform of a transparency illuminated by a collimated beam is obtained at the back focal plane of a lens. It, however, has a variable quadratic phase factor.
2. A pure Fourier transform (without the quadratic phase factor) is obtained when the transparency is placed at the front focal plane of the lens.

We have considered a very specific case of plane wave illumination of the transparency. When the transparency is illuminated by a spherical wave, it can be seen that the Fourier transform is obtained at a plane other than the focal plane. As mentioned earlier, the Fourier transform is obtained when Eq. (12) is satisfied. Equation (12) is also satisfied when

$$\frac{1}{d_0 + d_1} + \frac{1}{d_2} - \frac{1}{f} = 0 \tag{15}$$

This implies that a Fourier transform of an input is obtained at a plane that is conjugate to the point source plane. Indeed, a pure Fourier transform with unit scale factor is obtained when the transparency placed at the front focal plane is illuminated by a diverging spherical wave. Fourier transforms can be obtained by means of many other configurations, as well.

V. OPTICAL FILTERING

Availability of the Fourier transform at a physical plane leads one to think of ways and means to modify this by means of a transparency or a mask called a filter. In general, the Fourier transform of an object is complex: both the amplitude and the phase of the transform are to be operated upon by the filter. In some cases, certain frequencies from the spectrum are filtered out. Experiments of Abbe and also of Porter for the verification of Abbe's theory of microscope imaging are beautiful examples of optical filtering. Figure 3 shows a schematic of an optical filtering setup. This is the well-known $4f$ processor. The filter, either as a mask or a transparency, is placed at the filter plane where the Fourier transform of the input is displayed. A further Fourier transformation yields a filtered image.

Phase contrast microscopy exploits optical filtering where the zeroth order is phase advanced or retarded by $\pi/2$. The schlieren technique used in aerodynamics and combustion to visualize

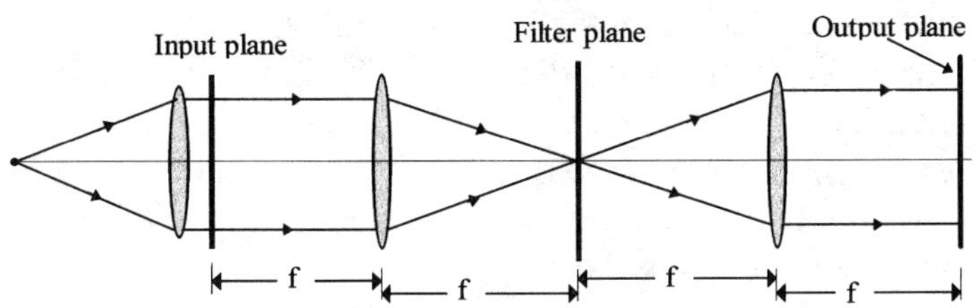

Figure 3 Optical filtering arrangement.

refractive index gradient is another example of optical filtering. Special masks have been used at the aperture stop in imaging lenses to alter its transfer function, hence to selectively filter the band of frequencies for image formation. Optical filtering is used in data processing, image processing, and pattern recognition. Theta modulation and frequency modulation are used to code the object so that certain features can be extracted by optical filtering.

VI. OPTICAL COMPONENTS IN OPTICAL METROLOGY

A light beam can be manipulated—it can be bent, dispersed, focused, and otherwise modified. The manipulation is done by a variety of optical components including lenses, mirrors, prisms, gratings, and their combinations. The action of a lens on an optical wave was discussed earlier. Since, however, the lens was assumed to be of infinite size, diffraction effects were neglected. In practice, a lens does not bring a plane wave to a point focus but to a finite distribution known as an Airy pattern. The intensity distribution in the image of a point source or at a focal plane when the plane wave is incident on it is given by $\left[2J_1(x)/x\right]^2$ where x is the argument of the Bessel function $J_1(..)$ and depends on the f number of the lens, the magnification, and the wavelength of the light. The image of a point is not a point but a distribution, which consists of a central disc where most of the energy is contained, surrounded by circular rings of decreasing intensity as shown in Figure 4. The radius of the Airy disc is given by $1.22\lambda f/2R$, where $2R$ is the lens aperture and f its focal length. It should be kept in mind that if the beam incident on the lens has a diameter smaller than the lens aperture, $2R$ must be replaced by the beam diameter. The intensity distribution is governed by the shape and size of the aperture.

If we carry out an analysis in spatial frequency domain, the lens is considered to be a low pass filter. The frequency response can be modified by apodization, as is the case with annular aperture telescopes. Use of an appropriate aperture mask on the lens alters its frequency response drastically.

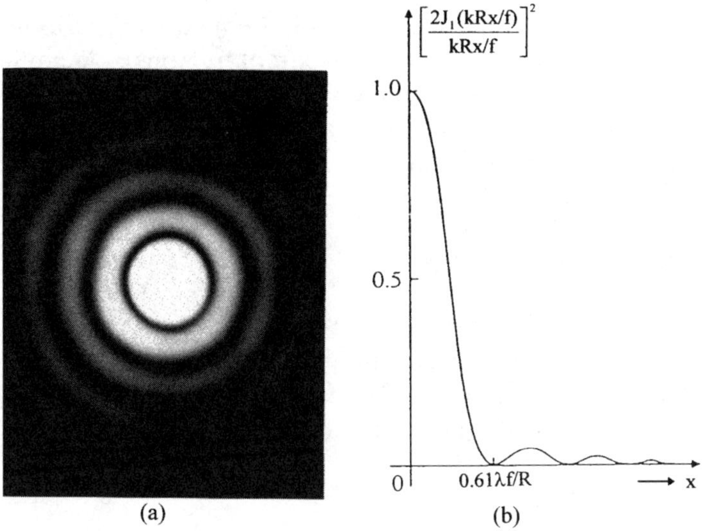

Figure 4 Diffraction at a lens of circular aperture: (a) photograph of the intensity distribution at the focus when a plane wave is incident on it and (b) plot of intensity distribution.

Similar arguments apply when the beam is manipulated by a mirror. A plane mirror merely changes the direction of the light beam. It always forms a virtual image free from aberrations. Spherical mirrors form both real and virtual images: a convex mirror always forms a reduced virtual image, while a concave mirror can form both real and virtual images of an object, depending on the object position and focal length. Mirror imaging is free from chromatic aberration. A prism, on the other hand, changes the direction of propagation of a monochromatic beam and disperses a nonmonochromatic beam. In optical metrology, a prism is mostly used to alter the direction of propagation—it can slightly tilt the beam as is done by a wedge prism, or change its direction by any arbitrary magnitude, say, 90° or 180°.

A grating is a dispersive element, hence is mainly the subject matter of spectrometers, spectrophotometers, and so on. A low frequency grating is used in moiré metrology. Highly accurate measurements are performed with gratings of fine pitch and monochromatic illumination. A grating is usually defined as a periodic arrangement of slits (opaque and transparent regions). A collimated beam incident on it is diffracted into a number of orders. When the grating is illuminated normally by a monochromatic beam of wavelength λ, the mth diffraction order is in the direction θ_m such that

$$d \sin \theta_m = m\lambda \qquad \text{for } m = 0, \pm 1, \pm 2, \pm 3, \ldots \tag{16}$$

where d is the pitch of the grating.

This grating will generate a large number of orders ($m \leq d/\lambda$, as $\theta_m \leq \pi/2$). Assuming small angle diffraction (i.e., diffraction from a coarse grating) and diffraction in the first order, we obtain

$$\theta = \frac{\lambda}{d} \qquad \text{or} \qquad \frac{1}{d} = \frac{\theta}{\lambda} \tag{17}$$

Suppose a lens of focal length f is placed behind the grating; this order will then be focused to a spot such that $x = \lambda f/d$ or $1/d = x/\lambda f$, where x is the distance of the spot from the optical axis. The grating is assumed to be oriented so that its grating elements are perpendicular to the x axis. We define the spatial frequency of the grating as the inverse of period d in millimeters. Thus, the spatial frequency $\mu = 1/d$ is

$$\mu = \frac{x}{\lambda f} \tag{18}$$

and is referred to as lines per millimeter (l/mm). Therefore, the spatial frequency μ and the off-axis distance x are linearly related for a coarse grating. In fact, the spatial frequency is defined as $\mu = \sin\theta/\lambda$. It is thus obvious that a sinusoidal grating of frequency μ will generate only three orders, $m = 0, 1, -1$.

A real grating can be represented as a summation of sinusoidal gratings whose frequencies are integral multiples of the fundamental. When this grating is illuminated by a collimated beam, a large number of spots are formed on either side of the axis: symmetric pairs correspond to a particular frequency in the grating.

If a grating of pitch p is inclined with respect to the x axis as shown in Figure 5, its spatial frequencies μ and ν along the x and the y directions are $\mu = 1/d_x$ and $\nu = 1/d_y$ such that

$$\frac{1}{\rho^2} = \frac{1}{d_x^2} + \frac{1}{d_y^2} \tag{19}$$

Alternately, $\rho^2 = \mu^2 + \nu^2$, where ρ is the spatial frequency of the grating. Diffraction at the inclined sinusoidal grating again results in the formation of three diffraction orders, which will lie on neither the x nor the y axis. It will be shown that the diffraction spots always lie on a line parallel to the grating vector.

A. Sinusoidal Grating

Interference between two inclined plane waves of equal amplitudes generates an interference pattern in which the intensity distribution is of the form

$$I(x) = 2I_0\left[1 + \cos\left(\frac{2\pi x}{d_x}\right)\right] \tag{20}$$

where d_x is the grating pitch and I_0 is the intensity of each beam. We associate a spatial frequency μ, which is the reciprocal of d_x in

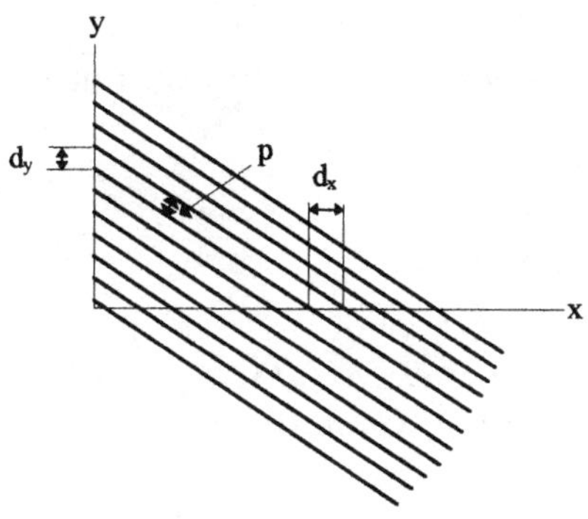

Figure 5 An inclined grating.

millimeters (i.e., $\mu = 1/d_x$). Thus, we can express the intensity distribution as follows:

$$I(x) = 2I_0[1 + \cos 2\pi\mu x] \tag{21}$$

Assuming an ideal recording material that maps the incident intensity distribution before exposure to amplitude transmittance after exposure, the amplitude transmittance of the grating record is given by

$$t(x) = 0.5[1 + \cos 2\pi\mu x] \tag{22}$$

The factor 0.5 appears because the transmittance varies between 0 and 1. Let a unit amplitude plane wave be incident normally on the grating. The amplitude just behind the grating is

$$a(x) = 0.5[1 + \cos 2\pi\mu x] = 0.5[1 + 0.5e^{2\pi i\mu x} + 0.5e^{-2\pi i\mu x}] \tag{23}$$

This represents a combination of three plane waves: one wave propagating along the axis and other two propagating inclined to the axis. A lens placed behind the grating will focus these plane waves to three diffraction spots, each spot corresponding to a plane wave. These spots lie on the x axis in this particular case.

Let us now recall the grating equation when a plane wave is incident normally, that is:

$$d_x \sin\theta_m = m\lambda$$

In the first order $m = 1$, and assuming small-angle diffraction, $\sin\theta \approx \theta$, we obtain

$$\frac{1}{d_x} = \mu = \frac{\theta}{\lambda}$$

From Figure 6, $\theta = x_f/f$, and hence $\mu = x_f/\lambda f$. Thus, since the position of the spot is directly proportional to the frequency of the grating, we have the implication that the diffraction at a sinusoidal grating generates three diffraction orders (the zeroth order lies on the optical axis and the first orders lie off-axis), whose displace-

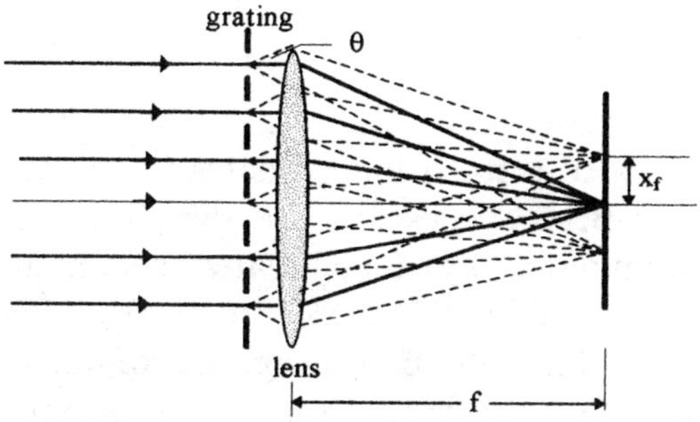

Figure 6 Diffraction at a grating.

ments are proportional to the spatial frequency of the grating. These orders lie on a line that is parallel to the grating vector $(2\pi/d_x)$. If the grating is rotated in its plane, these orders also rotate by the same angle.

In reality, a grating generates many spots. This can be easily observed by shining an unexpanded laser beam and observing the diffracted orders on a screen placed some distance away from the grating. This is because the grating transmittance differs significantly from the sinusoidal profile. Since a grating is defined as a periodic structure of transparent and opaque parts, it gives rise to infinity of orders subject to the condition that the diffraction angle $\theta \leq \pi/2$. It may be remembered that a linear relationship between spatial frequency and distance holds valid only under conditions of low angle diffraction or for coarse gratings.

B. Phase Grating

A phase grating has a periodic phase variation—it does not attenuate the beam. It can be realized either by thickness changes or by refractive index changes or their combination. One of the simplest methods is to record an interference pattern on a photoemulsion and then bleach it. Assuming that the phase variation is linearly related to the intensity distribution, the transmittance of a sinusoidal phase grating can be expressed as follows:

$$t(x) = \exp(i\phi_0)\exp[i\phi_m \cos(2\pi\mu_0 x)]$$

where ϕ_0 and ϕ_m are the constant phase and phase modulation, and μ_0 is the spatial frequency of the grating. This transmittance can be expressed in a Fourier–Bessel series. It is then implied that when such a grating is illuminated by a collimated beam, an infinity of orders is produced, unlike the three orders from an amplitude sinusoidal grating. The phase grating, therefore, is intrinsically nonlinear. The intensity distribution in the various orders can be controlled by groove shape, modulation, and also the frequency.

BIBLIOGRAPHY

1. A. Sommerfeld, *Optics*, Academic Press, New York (1964).
2. W. T. Cathey, *Optical Information Processing and Holography*, John Wiley, New York (1974).
3. Max Born and Emil Wolf, *Principles of Optics*, Pergamon Press, Oxford (1975).
4. J. D. Gaskill, *Linear Systems, Fourier Transforms and Optics*, John Wiley, New York (1978).
5. D. Casasent (Ed.), *Optical Data Processing*, Springer-Verlag, Berlin (1978).
6. S. H. Lee (Ed.), *Optical Data Processing—Fundamentals*, Springer-Verlag, Berlin (1981).
7. R. S. Sirohi, *Wave Optics and Applications*, Orient Longmans, Hyderabad (1993).
8. Joseph W. Goodman, *An Introduction to Fourier Optics*, McGraw-Hill, New York (1996).

Chapter 3

Phase Evaluation Methods

Phase distribution is encoded in an intensity distribution as a result of the phenomenon of interference. Phase distribution should therefore be retrieved from the interference pattern. We employ the interference phenomenon in classical interferometry, in holographic interferometry, and in speckle interferometry to convert the phase of a wave of interest into intensity distribution. Even in moiré methods, the desired information can be cast as an intensity distribution. The fringes in an interference pattern are loci of constant phase difference. Earlier methods of extracting phase information and consequently, the connected parameters of interest, were very laborious and time-consuming and suffered from the inherent inaccuracies of evaluation procedures. With the easy availability of computational and processing power and charge-coupled device (CCD) array detectors, many automatic fringe evaluation procedures have been developed.

There are many methods of phase evaluation. A comparison of these methods with respect to several parameters is presented in Table 1.

Table 1 Comparison of Some Prominent Methods of Phase Evaluation on Selected Parameters

Parameters	Methods			
	Fringe skeletonizing	Phase stepping and shifting	Fourier transform	Temporal heterodyning
Number of interferograms to be reconstructed	1	Minimum 3	1 (2)	One per detection point
Resolution (λ)	1–1/10	1/10–1/100	1/10–1/30	1/100–1/1000
Evaluation between intensity extrema	No	Yes	Yes	Yes
Inherent noise suppression	Partially	Yes	No (Yes)	Partially
Automatic sign detection	No	Yes	No (Yes)	Yes
Necessary experimental manipulation	No	Phase shift	No (phase shift)	Frequency
Experimental effort	Low	High	Low	Extremely high
Sensitivity to external influences	Low	Moderate	Low	Extremely high
Interaction by the operator	Possible	Not possible	Possible	Not possible
Speed of evaluation	Low	High	Low	Extremely low
Cost	Moderate	High	Moderate	Very high

We now discuss these methods in detail. During the course of this discussion, the significance of some of the parameters and characteristics of the various methods will become obvious.

I. INTERFERENCE EQUATION

Interference between two waves of the same frequency results in an intensity distribution of the form

$$I(x, y) = I_0(x, y)[1 + V(x, y) \cos \delta(x, y)] \quad (1)$$

where $I_0(x, y)$ is the low frequency background, often called the dc background, $V(x, y)$ is the fringe visibility (or modulation) and $\delta(x, y)$ is the phase difference between the two waves, which is to be determined. Here (x, y) represent the coordinates of a point, or of a pixel on an observation plane. The intensity distribution may be degraded by several factors (speckle noise, spurious fringe noise, electronic noise, etc.). However, when all these factors are included, the intensity distribution may still be expressed as follows:

$$I(x, y) = a(x, y) + b(x, y)\ \cos\delta(x, y) \tag{2}$$

This equation has three unknowns: $a(x, y)$, $b(x, y)$, and $\delta(x, y)$. We are mostly interested in obtaining $\delta(x, y)$ and hence will discuss the methods of phase evaluation only.

II. FRINGE SKELETONIZATION

The fringe skeletonizing methods are computerized forms of the manual fringe counting methods used formerly. The fringe pattern is received on a CCD detector. It is assumed that the local extremum of the intensity distribution corresponds to the maximum or minimum of the cosine function. (In speckle photography fringe patterns, this requirement is not met.) In this case, the interference phase at pixels where an intensity maximum or minimum is detected is an even or odd integer multiple of π. Therefore, the fringe skeletonization seeks maxima or minima of an intensity distribution; the fringe locations are thus determined within several pixels. The method is thus similar to the manual method of locating the fringe maxima or minima but is performed much faster and with relatively high accuracy. The sign ambiguity is still not resolved.

The methods for fringe skeletonization can be divided into those based on fringe tracking and those related to segmentation. Both techniques require careful preprocessing to minimize speckle noise through low pass filtering and to correct for an uneven brightness distribution in the background illumination.

III. TEMPORAL HETERODYNING

In temporal heterodyning, a small frequency difference, $\Delta\omega/2\pi = [(\omega_1 - \omega_2)/2\pi]$, which is less than 100 kHz, is introduced between the two interfering waves. The local intensity of the interference pattern then varies sinusoidally at the beat frequency, $\Delta\omega/2\pi$. The intensity distribution can be expressed as follows:

$$I(x, y : t) = a(x, y) + b(x, y) \cos[\Delta\omega t + \delta(x, y)] \tag{3}$$

There is no stationary stable interference pattern like the one familiar in interferometry; instead, the intensity at each point (x, y) varies sinusoidally at the beat frequency, and the interference phase at that point is transformed into the phase of the beat frequency signal. There is no way to measure the phase of this beat signal. Clearly, however, although the intensity at all points varies with the beat frequency, the phases are different, being equal to the values of the interference phases at these points. Since the beat frequency is low enough, the phase difference between two points can be measured with very high accuracy independently of $a(x, y)$ and $b(x, y)$ by using two photodetectors and an electronic phasemeter. In this way, both the interpolation problem and the sign ambiguity of classical interferometry are solved. The method requires special equipment such as acoustooptic modulators for frequency shift and a phasemeter. The image (interference pattern) is scanned mechanically by photodetectors to measure the phase difference: one detector is fixed and other scanned, thereby allowing for measurement of the interference phase with respect to the fixed photodetector. The measurement speed is low (typically one second per point) but the accuracy (typically $\lambda/1000$) and the spatial resolution ($> 10^6$ resolvable points) are extremely high.

In double-exposure holographic interferometry, both the interfering waves are simultaneously released from the hologram; hence temporal heterodyning cannot be applied for phase evaluation. On the other hand, if a two-reference-wave setup is used, so that each state of the object is recorded with one of the reference waves, this technique can be implemented (see Chapter 5). In practice, two acoustooptic modulators in cascade are placed in the path of one of the reference waves. These modulators introduce frequency shifts of opposite signs. During recording, both the modulators are driven at the same frequency, say, at 40 MHz; hence the net frequency shift is zero. The initial and final states of the object are recorded sequentially with one reference wave at a time. The reconstruction, however, is performed with both the reference waves simultaneously, with one modulator driven at 40 MHz and the other at 40.1 MHz. Therefore, the reference waves are of different frequencies, with a frequency difference of 100 kHz, resulting in an interference pattern that oscillates at that frequency.

One way to evaluate the interference pattern is to keep one detector fixed at the reference point while the other is scanned over the image. This way, one measures the phase difference with respect to the phase of the reference point. The phase differences are stored, arranged, and displayed. Instead of two detectors, one could use an array of three, four, or five detectors to scan the whole image. In this way, one measures the phase differences $\delta\delta_x$ and $\delta\delta_y$ between adjacent points along the x and the y directions with known separations. Numerical integration of recorded and stored phase difference values gives the interference phase distribution.

IV. PHASE SAMPLING EVALUATION: QUASI-HETERODYNING

As stated earlier, the intensity distribution in an interference pattern [Eq. (2)], has three unknowns; $a(x, y)$, $b(x, y)$, and $\delta(x, y)$. Therefore, if three intensity values are available at each point, we can set up three equations and solve them for the unknowns. These three intensity values are obtained by changing the phase of the reference wave. To be general, we set up n equations.

Two different approaches are available for quasi-heterodyning phase measurement, namely, the phase step method and the phase shift method. In the phase step method, the local intensity $I(x, y)$ in the interference pattern is sampled at fixed phases α_n of the reference wave, that is,

$$I_n(x, y) = a(x, y) + b(x, y) \cos\{\delta(x, y) + \alpha_n\} \quad \text{for } n = 1, 2, 3, \ldots, N \text{ but } \geq 3 \tag{4}$$

As mentioned earlier, at least three intensity measurements, I_1, I_2, I_3, must be carried out to determine all three unknowns.

In the phase-shifting or integrating bucket method, which is intended primarily for use with CCD detectors on which optical power is integrated by the detector, the phase of the reference wave is varied linearly and the sampled intensity is integrated over the phase interval $\Delta\alpha$ from $(\alpha_n - \Delta\alpha/2)$ to $(\alpha_n + \Delta\alpha/2)$. The intensity $I_n(x, y)$ sampled at a pixel at (x, y) is given by

$$\begin{aligned} I_n(x, y) &= \frac{1}{\Delta\alpha} \int_{\alpha_n - \Delta\alpha/2}^{\alpha_n + \Delta\alpha/2} \{a(x, y) + b(x, y) \cos[\delta(x, y) + \alpha(t)]\} \, d\alpha(t) \\ &= a(x, y) + \operatorname{sinc}\left(\frac{\Delta\alpha}{2}\right) b(x, y) \cos[\delta(x, y) + \alpha_n] \end{aligned} \tag{5}$$

This expression is equivalent to that of the phase-shifting method, the only difference being that the modulation $b(x, y)$ is modified; there is a reduction in the contrast of fringes by $\operatorname{sinc}(\Delta\alpha/2)$. In this sense, the phase-shifting method is equivalent to the phase-stepping method and the names are used synonymously. Furthermore, for data processing both methods are handled in an identical manner.

V. THE PHASE-SHIFTING METHOD

For the solution of a nonlinear system of equations, we use the Gaussian least-squares approach and rewrite the intensity distribution as follows:

$$\begin{aligned} I_n(x, y) &= a(x, y) + b(x, y) \cos[\delta(x, y) + \alpha_n] \\ &= a + u \cos\alpha_n + v \sin\alpha_n \end{aligned} \tag{6}$$

where $u(x, y) = b(x, y) \cos\delta(x, y)$, and $v(x, y) = -b(x, y) \sin\delta(x, y)$. We obtain a, u, and v by minimizing the errors. That is, we minimize the sum of the quadratic errors:

$$\sum_{n=1}^{N}[I_n(x, y) - (a + u\cos\alpha_n + v\sin\alpha_n)]^2$$

Partially differentiating this function with respect to a, u, and v and then equating the derivatives to zero gives a linear system of three equations:

$$\begin{pmatrix} N & \sum\cos\alpha_n & \sum\sin\alpha_n \\ \sum\cos\alpha_n & \sum\cos^2\alpha_n & \sum\sin\alpha_n\cos\alpha_n \\ \sum\sin\alpha_n & \sum\cos\alpha_n\sin\alpha_n & \sum\sin^2\alpha_n \end{pmatrix}\begin{pmatrix} a \\ u \\ v \end{pmatrix} = \begin{pmatrix} \sum I_n \\ \sum I_n\cos\alpha_n \\ \sum I_n\sin\alpha_n \end{pmatrix} \tag{7}$$

This system is to be solved pointwise. We obtain

$$a(x, y) = \frac{1}{N}\sum_{n=1}^{N} I_n \tag{8a}$$

$$b(x, y) = \frac{[(\sum I_n\cos\alpha_n)^2 + (\sum I_n\sin\alpha_n)^2]^{1/2}}{N} \tag{8b}$$

and

$$\tan\delta(x, y) = -\frac{v}{u} = -\frac{\sum I_n\sin\alpha_n}{\sum I_n\cos\alpha_n} \tag{8c}$$

The interference phase is computed modulo 2π. A large number of algorithms have been mentioned in the literature. Table 2 presents some of these algorithms, derived from the interference equation, using three and four steps.

Table 2 Some Algorithms Derived from the Interference Equation

N	Phase step α_n	Expression for $\tan\delta(x, y)$
3	$60°(0°, 60°, 120°)$	$\tan\delta(x, y) = \dfrac{2I_1 - 3I_2 + I_3}{\sqrt{3}(I_2 - I_3)}$
3	$90°(0°, 90°, 180°)$	$\tan\delta(x, y) = \dfrac{I_1 - 2I_2 + I_3}{I_1 - I_3}$
3	$120°(0°, 120°, 240°)$	$\tan\delta(x, y) = \dfrac{\sqrt{3}(I_3 - I_2)}{2I_1 - I_2 - I_3}$
4	$90°(0°, 90°, 180°, 270°)$	$\tan\delta(x, y) = \dfrac{I_4 - I_2}{I_1 - I_3}$
4	$60°(0°, 60°, 120°, 180°)$	$\tan\delta(x, y) = \dfrac{5(I_1 - I_2 - I_3 + I_4)}{\sqrt{3}(2I_1 + I_2 - I_3 - 2I_4)}$
5	$90°(0°, 90°, 180°, 270°, 360°)$	$\tan\delta(x, y) = \dfrac{7(I_4 - I_2)}{4I_1 - I_2 - 6I_3 - I_4 + 4I_5}$

VI. PHASE SHIFTING WITH UNKNOWN BUT CONSTANT PHASE STEP

There exists a method known as the Carré method, which utilizes four phase-shifted intensity distributions; the phase step need not be known, but it must remain constant during phase shifting. The four intensity distributions are expressed as follows:

$$I_1(x, y) = a(x, y) + b(x, y)\ \cos[\delta(x, y) - 3\alpha] \tag{9a}$$

$$I_2(x, y) = a(x, y) + b(x, y)\ \cos[\delta(x, y) - \alpha] \tag{9b}$$

$$I_3(x, y) = a(x, y) + b(x, y)\ \cos[\delta(x, y) + \alpha] \tag{9c}$$

$$I_4(x, y) = a(x, y) + b(x, y)\ \cos[\delta(x, y) + 3\alpha] \tag{9d}$$

For the sake of simplicity, the phase step is taken as 2α. There are four unknowns, namely, $a(x, y)$, $b(x, y)$, $\delta(x, y)$, and 2α; hence it should be possible to obtain the values of all four unknowns from the four equations. We, however, present the expressions for the phase step and interference phase as follows:

$$\tan^2\alpha = \frac{3(I_2 - I_3) - (I_1 - I_4)}{(I_2 - I_3) + (I_1 - I_4)} \tag{10a}$$

and

$$\tan\delta(x, y) = \tan\alpha\,\frac{(I_2 - I_3) + (I_1 - I_4)}{(I_2 + I_3) - (I_1 + I_4)} \tag{10b}$$

These two equations are combined to yield the interference phase from the measured intensity values as follows:

$$\tan\delta(x, y) = \frac{\{[3(I_2 - I_3) - (I_1 - I_4)]\,[(I_2 - I_3) + (I_1 - I_4)]\}^{1/2}}{(I_2 + I_3) - (I_1 + I_4)} \tag{11}$$

An improvement over the Carrè method is realized by calculating the value of the phase step over the whole pixel format and then taking the average value for calculation. Constancy of the phase step is also easily checked. The phase step 2α is obtained from the expression

$$\cos 2\alpha = \frac{(I_1 - I_2) + (I_3 - I_4)}{2(I_2 - I_3)} \tag{12}$$

By assumption, the phase step 2α must be constant over all the pixels. We, however, take an average value that smooths out any fluctuations. The interference phase is then obtained as follows:

$$\tan \delta(x, y) = \frac{(I_3 - I_2) + (I_1 - I_3) \cos \bar{\alpha}_1 + (I_2 - I_1) \cos 2\bar{\alpha}_1}{(I_1 - I_3) \sin \bar{\alpha}_1 + (I_2 - I_1) \sin 2\bar{\alpha}_1} \tag{13a}$$

and

$$\tan \delta(x, y) = \frac{(I_4 - I_3) + (I_2 - I_4) \cos \bar{\alpha}_1 + (I_3 - I_2) \cos 2\bar{\alpha}_1}{(I_2 - I_4) \sin \bar{\alpha}_1 + (I_3 - I_2) \sin 2\bar{\alpha}_1} \tag{13b}$$

where $2\bar{\alpha}_1$ is the average value of the phase step 2α, averaged over all the pixels. The 2π steps of the interference phase distribution modulo 2π occur at different points in the image. This information can be used in the subsequent demodulation by considering the continuous phase variation of the two terms at each pixel.

There is another interesting algorithm that also utilizes four intensity distributions but can be used for the evaluation of a very noisy interferogram. We consider an intensity distribution in the image of an object being studied via speckle techniques. The intensity distribution is expressed as follows:

$$I(x, y) = a(x, y) + b(x, y) \cos(\phi_o - \phi_R), \tag{14}$$

where ϕ_o and ϕ_R are the phases of the speckled object wave and the reference wave, respectively; the phase difference $\phi(= \phi_o - \phi_R)$ is random. We capture four frames, the intensity distributions being expressed as follows:

$$I_1(x, y) = a(x, y) + b(x, y) \cos\left(\phi - \frac{\pi}{2}\right) \tag{15a}$$

$$I_2(x, y) = a(x, y) + b(x, y) \cos(\phi) \tag{15b}$$

$$I_3(x, y) = a(x, y) + b(x, y) \cos[\phi + \delta(x, y)] \tag{15c}$$

$$I_4(x, y) = a(x, y) + b(x, y) \cos\left[\phi + \delta(x, y) + \frac{\pi}{2}\right] \tag{15d}$$

Here $\delta(x, y)$ is the phase introduced by deformation. From these equations, we obtain

$$I_3 - I_2 = -2b \sin\left(\phi + \frac{\delta}{2}\right) \sin\left(\frac{\delta}{2}\right) \tag{16a}$$

and

$$I_4 - I_1 = -2b \sin\left(\phi + \frac{\delta}{2}\right) \cos\left(\frac{\delta}{2}\right) \tag{16b}$$

From Eqs. (16a) and (16b), the deformation phase is obtained easily as follows:

$$\tan\left[\frac{\delta(x, y)}{2}\right] = \frac{I_3 - I_2}{I_4 - I_1} \tag{17}$$

The phase is obtained modulo 2π as in other algorithms. However, the phase shift must be exactly $\pi/2$.

Phase-shifting methods are commonly used in interferometers of all kinds. The phase shift is usually introduced by calibrated PZT shifters attached to a mirror. This mechanical means of shifting the phase is quite common and convenient to use. There are various other methods, which rely on polarization and refractive index changes of the medium by electrical field, and so on.

Good features of the phase-shifting technique include (1) almost real-time processing of interferograms, (2) removal of sign ambiguity, (3) insensitivity to source intensity fluctuations, (4) good accuracy even in areas of very low fringe visibility, (5) high measurement accuracy and repeatability, and (6) applicability to holographic interferometry, speckle interferometry, moiré, and other methods. Further, we can calculate the visibility and background intensity from the measured intensity values.

VII. SPATIAL PHASE SHIFTING

Temporal phase shifting, as described above, requires the interference pattern to be stable over the period of phase shifting. When a transient event is being studied or the environment is highly turbulent, however, the results from temporal phase shifting may be erroneous. To overcome this problem, phase-shifted interferograms can be captured simultaneously by means of multiple CCD array detectors and appropriate splitting schemes. This, however, adds to the cost and complexity of the system. Another solution is to use spatial phase shifting. Spatial phase shifting requires an appropriate carrier. The intensity distribution in the interferogram can be expressed as follows:

$$I(x, y) = a(x, y) + b(x, y)\cos[2\pi f_0 x + \phi(x, y)] \tag{18}$$

where f_0 is the spatial carrier along the x direction. The spatial carrier is so chosen as to have either three or four pixels on the adjacent fringes to apply three-step or four-step algorithms. There are several ways to generate the carrier frequency. The simplest one is to tilt the reference beam appropriately. Polarization coding and gratings are also frequently used. We describe a procedure in speckle metrology later (Chapter 6).

We consider the experimental setup shown in Figure 1. This arrangement is employed for the measurement of the in-plane displacement component. The aperture separation is chosen so that

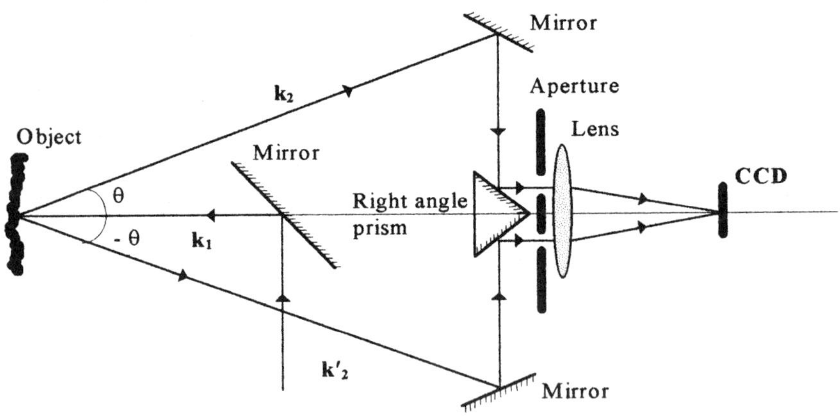

Figure 1 Measurement of in-plane component using spatial phase shifting.

the fringe width is equal to the width of three pixels of the CCD camera for the 120° phase shift algorithm and four pixel widths for the 90° phase shift algorithm as shown in Figure 2. The average speckle size is nearly equal to the fringe width. The intensity distribution $I(x_n, y)$ at the nth pixel along the x direction is given by

$$I(x_n, y) = a(x_n, y) + b(x_n, y)\,\mathrm{sinc}\left(\frac{\Phi}{2}\right)\cos[\phi(x_n, y) + n\beta + C]$$
$$\text{for } n = 1, 2, 3, \ldots, N \tag{19}$$

where $\phi(x_n, y)$ is the phase to be measured, $\beta = (2\pi/\lambda)(d/v)d_p$ is the phase change from one pixel to another, separated by d_p along the x direction, $\Phi = (2\pi/\lambda)(d/v)d_t$ is the phase shift angle over which the pixel of width d_t integrates the intensity, C is a constant phase offset, and N is the number of pixels in a row. Object deformation causes the initial phase $\phi_1(x_n, y)$ to change to $\phi_2(x_n, y) = \phi_1(x_n, y) + \delta(x_n, y)$. If $\phi_1(x_n, y)$ and $\phi_2(x_n, y)$ are measured, the phase difference $\delta(x_n, y)$ can be obtained. To calculate $\phi_n = \phi(x_n, y)$, we need a minimum of three adjacent values of $I(x_n, y)$. If these values are picked up at 120° phase intervals, then the phase ϕ_n is given by

$$\phi_n = \tan^{-1}\left[\sqrt{3}\,\frac{I_{n-1} - I_{n+1}}{2I_n - I_{n-1} - I_{n+1}}\right] \bmod \pi \tag{20}$$

and

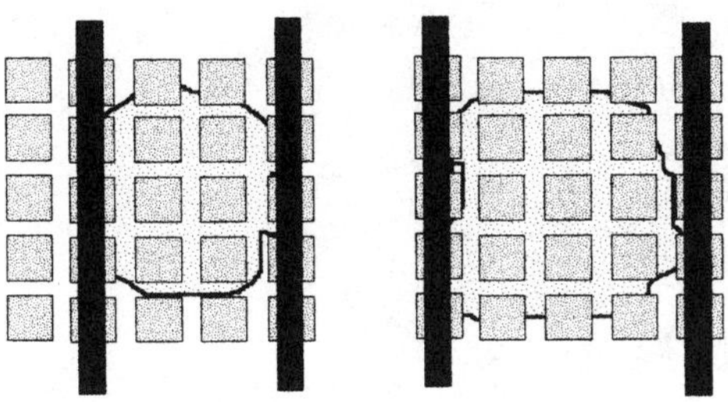

(a) (b)

Figure 2 (a) 120° spatial phase shifting, and (b) 90° spatial phase shifting.

$$I_{n+m} = a + b\,\mathrm{sinc}\left(\frac{\Phi}{2}\right)\cos\left[\phi_n + \frac{2\pi}{\lambda}\frac{d}{v}\,d_p(n+m)\right] \tag{21}$$
$$\text{for } n = 2,\ 3, \ldots, (N-1);\quad m = -1,\ 0,\ +1$$

The constant C has been dropped from Eq. (21). It may be seen that in spatial phase shifting, the modified phase $\psi_n = \phi_n + (2\pi/\lambda)(d/v)d_p n$ (mod π), rather than the speckle phase ϕ_n (mod π), is reconstructed. Therefore, the phase offset $(2\pi/\lambda)(d/v)d_p n$ must be subtracted in a later step.

VIII. THE FOURIER TRANSFORM METHOD

The Fourier transform (FT) method is used in two ways: with a spatial frequency carrier added to it (spatial heterodyning) and without the carrier frequency. We will first describe the FT method without the spatial carrier. We express the intensity distribution in the interference pattern again as follows:

$$I(x, y) = a(x, y) + b(x, y)\cos\delta(x, y)$$

With the cosine function written in complex exponentials, that is,

$$2\cos\delta(x, y) = e^{i\delta(x,y)} + e^{-i\delta(x,y)}$$

the intensity distribution can be expressed as follows:

$$\begin{aligned} I(x, y) &= a(x, y) + 0.5b(x, y)e^{i\delta(x,y)} + 0.5b(x, y)e^{-i\delta(x,y)} \\ &= a(x, y) + c(x, y) + c^*(x, y) \end{aligned} \tag{22}$$

where $c(x, y)$ and $c^*(x, y)$ are now complex functions. We take the Fourier transform of this distribution, yielding

$$\mathbf{I}(\mu, \nu) = A(\mu, \nu) + C(\mu, \nu) + C^*(-\mu, -\nu) \tag{23}$$

with μ and ν being the spatial frequencies. Since the intensity distribution $I(x, y)$ is a real-valued function in the spatial domain, its Fourier transform is Hermitian in the frequency domain, that is,

$$\mathbf{I}(\mu, \nu) = \mathbf{I}^*(-\mu, -\nu) \tag{24}$$

The real part of $\mathbf{I}(\mu, \nu)$ is even and the imaginary part is odd. The amplitude distribution $|\mathbf{I}(\mu, \nu)|$ is point-symmetric with respect to the dc term $\mathbf{I}(0, 0)$. Now it is seen that $A(\mu, \nu)$ contains the zero peak $\mathbf{I}(0, 0)$ and the low frequency content due to slow variation of the background. $C(\mu, \nu)$ and $C^*(-\mu, -\nu)$ carry the same information, but with sign ambiguity. By bandpass filtering in the spatial frequency domain, $A(\mu, \nu)$ and either $C(\mu, \nu)$ or $C^*(-\mu, -\nu)$ is filtered out. The remaining spectrum $C^*(-\mu, -\nu)$ or $C(\mu, \nu)$ is not Hermitan, so the inverse Fourier transform gives complex $c(x, y)$ with nonvanishing real and imaginary parts. The interference phase $\delta(x, y)$ is obtained as follows:

$$\delta(x, y) = \tan^{-1} \frac{\mathrm{Im}[c(x, y)]}{\mathrm{Re}[c(x, y)]} \tag{25}$$

where Re [ ] and Im [ ] represent real and imaginary parts, respectively.

IX. SPATIAL HETERODYNING

In spatial heterodyning, a carrier frequency is added to the interference pattern. The spatial frequency is chosen higher than the maximum frequency content in $a(x, y)$, $b(x, y)$, and $\delta(x, y)$. The intensity distribution in the interference pattern can then be expressed as follows:

$$I(x, y) = a(x, y) + b(x, y)\ \cos\{\delta(x, y) + 2\pi f_0 x\}$$

where f_0 is the spatial carrier frequency along the x direction. In interferometry, the carrier is usually added by tilting the reference mirror. There are other ways of introducing the carrier in the holographic, speckle, and electronic speckle pattern modes of interferometry. The Fourier transform of this expression can be expressed as follows:

$$\mathbf{I}(\mu, \nu) = A(\mu, \nu) + C(\mu - f_0, \nu) + C^*(-\mu - f_0, -\nu) \tag{26}$$

Since the spatial carrier frequency is chosen higher than the maximum frequency content in $a(x, y)$, $b(x, y)$, and $\delta(x, y)$, the spectra $A(\mu, \nu)$, $C(\mu - f_0, \nu)$, and $C^*(-\mu - f_0, -\nu)$ are separated. The spectrum $A(\mu, \nu)$ is centered on $\mu = 0$, $\nu = 0$ and carries the information about the background. The spectra $C(\mu - f_0, \nu)$ and $C^*(-\mu - f_0, -\nu)$ are placed at $\mu = f_0$, $\nu = 0$ and $\mu = -f_0$, $\nu = 0$ (i.e., symmetrically about the dc term). If, by means of an appropriate bandpass filter, $A(\mu, \nu)$ and $C^*(-\mu - f_0, \nu)$ are eliminated and subsequently $C(\mu - f_0, \nu)$ is shifted by f_0 toward the origin, thereby removing the carrier, we can obtain $c(x, y)$ by inverse Fourier transformation. The interference phase is then obtained from the real and imaginary parts of $c(x, y)$.

If the inverse transform of $C^*(-\mu, -\nu)$ instead of $C(\mu, \nu)$ were taken, it would result in $-\delta(x, y)$. The sign ambiguity is always present when a single interferogram is evaluated. The information about the sign of the phase is obtained when an additional phase-shifted interferogram is available. Let us now write the expressions for the intensity distributions in the two interferograms, with one of them phase-stepped by α, as follows:

$$I_1(x, y) = a(x, y) + b(x, y) \cos \delta(x, y)$$

$$I_2(x, y) = a(x, y) + b(x, y) \cos[\delta(x, y) + \alpha]$$

Theoretically, the value of α must be in the range $0 < \alpha < \pi$; in practice, a value in the range $\pi/3 < \alpha < 2\pi/3$ is recommended. If this condition is fulfilled, the exact value of α need not be known. Again, we can express these intensity distributions in terms of complex exponentials and then take the Fourier transforms, as described earlier. After bandpass filtering and taking the inverse Fourier transforms of the spectra belonging to each intensity distribution, we will obtain

$$c_1(x, y) = 0.5b(x, y)e^{i\delta(x,y)}$$
$$c_2(x, y) = 0.5b(x, y)e^{i\{\delta(x,y)+\alpha\}}$$

The phase step α is calculated pointwise from the expression

$$\alpha(x, y) = \tan^{-1} \frac{\text{Re}[c_1(x, y)] \, \text{Im}[c_2(x, y)] - \text{Im}[c_1(x, y)] \, \text{Re}[c_2(x, y)]}{\text{Re}[c_1(x, y)] \, \text{Re}[c_2(x, y)] + \text{Im}[c_1(x, y)] \, \text{Im}[c_2(x, y)]} \tag{27}$$

The knowledge of $\alpha(x, y)$ is used for the determination of the sign-corrected interference phase distribution $\delta(x, y)$ from the expression

$$\delta(x, y) = \text{sign}\{\alpha(x, y)\} \tan^{-1} \frac{\text{Im}[c_1(x, y)]}{\text{Re}[c_1(x, y)]} \tag{28}$$

The phase is unwrapped as usual.

BIBLIOGRAPHY

1. E. Wolf (Ed.), *Progress in Optics*, Vol. 26, Elsevier Science, Amsterdam (1988).
2. E. Wolf (Ed.), *Progress in Optics*, Vol. 28, Elsevier Science, Amsterdam (1990).
3. Wolfgang Osten, *Digital Processing and Evaluation of Interference Images*, Akademie Verlag, Berlin (1991).
4. David W. Robinson and Graeme T. Reid (Eds.), *Interferogram Analysis—Digital Fringe Measurement Techniques*, Institute of Physics Publishing, Bristol (1993).
5. Thomas Kreis, *Holographic Interferometry, Principles and Methods*, Akademie Verlag Series in Optical Metrology, Vol. 1, Akademie Verlag, Berlin (1996).
6. Pramod K. Rastogi (Ed.), *Optical Measurement Techniques and Applications*, Artech House, Boston (1997).

Chapter 4

Detectors and Recording Materials

Detection is a very important step in the measurement process. A detector converts incident optical energy into electrical energy, which is then measured. Among the many physical effects that have been used to detect optical radiation are photoexcitation and photoemission, photoresistive effect, and photothermal effect. Detectors based on photoexcitation or photoemission of electrons are by far the most sensitive detectors and provide the highest performance. These detectors are, however, small-area detectors, and they measure or sample the optical energy over a very small area. In techniques and methods described in this book, large-area or image detectors are employed except for heterodyne HI. These include photographic plates, photochromic devices, thermoplastics, and photorefractive crystals. The CCD arrays now often used consist of an array of small-area (pixel) detectors based on photoexcitation. We therefore first discuss the semiconductor devices as detectors, then photomultiplier tubes (PMTs), and finally the image detectors.

A variety of detectors are used to detect and/or measure optical signals. Photodetectors fall into two general categories: thermal detectors and quantum detectors. In thermal detectors, the incident thermal energy is converted to heat, which in turn may cause a rise in temperature and a corresponding measurable change such as an emf. These detectors are not used in this book, hence are not discussed. Examples of quantum detectors are photomultiplier tubes, semiconductor photodiodes, photoconductors, and phototransistors. In this type of detector, the incident optical energy (i.e., the photons) produces emission of electrons in PMTs, or generation of electron–hole pairs in diodes. A resistance change is directly produced in photoconductors by the absorption of photons.

The detectors are compared by the use of several defined parameters: responsivity R, detectivity D or D^*, NEP (noise equivalent power), noise, spectral response, and frequency response. The responsivity is the ratio of the output signal (in amperes or volts) to the incident intensity and is given in microamperes per milliwatt per square centimeter ($\mu A/mW/cm^2$). For photodiodes, the quantum efficiency, which is the number of electron–hole pairs per incident photon, is sometimes given. Responsivity is proportional to quantum efficiency. The detectivity is the ratio of the responsivity to the noise current (voltage) produced by the detector. This, therefore, is the signal-to-noise (S/N) ratio divided by intensity. An often used parameter is D^*, which includes the dependence on noise frequency bandwidth Δf and detector area A and is connected to the detectivity through the relation

$$D^* = D(A\Delta f)^{1/2} \tag{1}$$

NEP is the reciprocal of D. It is thus the light intensity required to produce a signal-to-noise ratio of 1. There are several sources of noise [Johnson (thermal) noise, shot noise, generation–recombination noise, etc.]. In all these cases, the noise current is proportional to $(\Delta f)^{1/2}$. Spectral response refers to the variation of responsivity as a function of wavelength. Finally, the frequency response of a detector refers to its ability to respond to a chopped or modulated beam. Most solid state detectors behave like low pass filters, with the responsivity given as follows:

$$R = R_0 \frac{1}{\left(1 + \omega^2 \tau^2\right)^{1/2}} \tag{2}$$

where R_0 is the responsivity at zero frequency, $f(= \omega/2\pi)$ is the frequency of modulation, and τ is the time constant. The cutoff frequency f_c is $f_c = (2\pi\tau)^{-1}$, where $R = R_0/\sqrt{2}$.

I. DETECTORS

The incident photon creates an electron–hole pair in semiconductor devices and release of an electron in a PMT. There is thus a frequency below which the detector does not respond to the optical energy. The frequency ν of the radiation must be greater than the threshold frequency ν_{th} where $h\nu_{th} = E$, E being the band gap in semiconductor or the work function of the cathode material in PMT; h is the Planck's constant.

A. Photoconductors

Photoconductor detectors are of two types: intrinsic and extrinsic. In an intrinsic semiconductor, an incident photon, upon absorption, excites an electron from the valence band to the conduction band. This requires that the photon energy $h\nu \gg E_g$, where E_g is the band gap. The cutoff wavelength λ_c is given by

$$\lambda_c = \frac{1.24}{E_g} \tag{3}$$

where λ_c is in micrometers and E_g in electron-volts. The spectral response of the detector can be tailored by the choice of appropriate E_g. In fact, the absorption of photons results in the generation of both excess electrons and holes, which are free to conduct under an applied electric field. The excess conductivity is called the intrinsic photoconductivity.

When the incident photons excite the electrons from donor levels to the conduction band or holes from acceptor levels to the valence band, the excess conductivity is called the extrinsic photoconductivity. In this case, excitation creates only one type of excess carriers, electrons or holes, by ionizing a donor or an acceptor, respectively. The spectral response is determined by the donor or the acceptor energies. For donor excitation, the cutoff wavelength is given by

$$\lambda_c = \frac{1.24}{E_c - E_d} \tag{4}$$

where E_c and E_d are the energies of the conduction band and donor level, respectively. Since $(E_c - E_d)$ or $(E_a - E_v)$ is much smaller than E_g, these detectors are sensitive in long-wavelength IR region.

When a beam of light of intensity I and frequency ν is incident on a photoconductor detector, the rate of generation of carriers can be expressed as follows:

$$G = \eta \frac{I}{h\nu} \tag{5}$$

where η is the quantum efficiency. The photocurrent i generated in the external circuit is

$$i = \eta \frac{e}{h\nu} \left(\frac{\tau_0}{\tau_d} \right) I \tag{6}$$

where τ_0 is the lifetime of the carriers, τ_d is the transit time (time required by the carriers to traverse the device), and the factor (τ_0/τ_d) is interpreted as photoconductive gain. It is seen that the photocurrent is proportional to the light intensity.

Photoconductors exhibit the memory effect; that is, at a given illumination, the photoconductor may have several resistance values depending on the device's history of illumination. These devices are not stable, hence are not suited for precise and accurate measurements. They are, however, inexpensive, simple, bulky and durable.

B. Photodiodes

The commonly known photodiodes include the simple p-n junction diode, the p-i-n diode, and the avalanche diode. The fabrication and operation of the photodiodes are based on p-n technology. A p-n junction is formed by bringing p-type and n-type materials together. The electrons are dominant charge carriers in n-type semiconductor, whereas holes are dominant in p-type. At the instant when the materials are joined together, there is an almost infinite concentration gradient across the junction for both the electrons and the holes (Fig. 1a). Consequently, the electrons and the holes diffuse in opposite directions. The diffusion process, however, does not continue indefinitely; rather, a potential barrier is developed, which opposes the diffusion. For every free electron leaving an n-type region, an immobile positive charge is left behind. The amount of positive charge increases as the number of departing electrons increases. Similarly, as the holes depart from the p-type region, immobile negative charges build up in the p-type region. The mobile electrons and holes combine and neutralize each other, and form on both sides of the junction a region devoid of free charge carriers. This region is called the depletion region, and it acts as an insulator owing to the absence of free carriers. The immobile charges of opposite polarity generate a barrier voltage that opposes further diffusion (Fig. 1b). The barrier voltage depends on the semiconductor material. Its value for silicon is 0.7 V. The junction appears like a capacitor (i.e., a nonconducting material separating the two conductors).

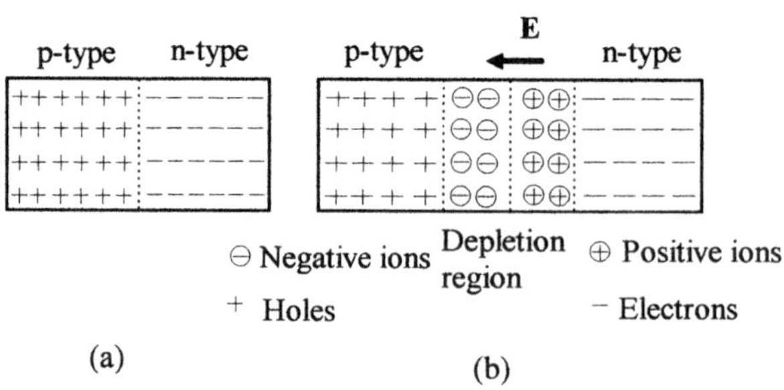

Figure 1 Two p-n junctions: (a) just formed and (b) on equilibrium.

When forward bias is applied to the junction, it opposes the barrier voltage, thereby reducing the thickness of the depletion region and increasing the junction capacitance. When the bias reaches the barrier voltage, the depletion region is eliminated and the junction becomes conductive. When the reverse bias is applied, the depletion region is widened, the junction capacitance is reduced, and the junction stays nonconductive. The reverse bias junction, however, can conduct current when free carriers are introduced into the junction. These free carriers can be introduced by an incident radiation. When the radiation falls in the depletion region or within the diffusion length around it, electron–hole pairs are created. The electrons (as minority carriers in a p-type region) will drift toward the depletion region, cross it, and hence contribute to the external current. Similarly, holes created by photoabsorption in the n-type region will drift in the opposite direction and will contribute to the current. Within the depletion region, however, the electron–hole pairs will be separated; electrons and holes will then drift from each other in opposite directions, and thus both will contribute to the current flow.

If we consider only the photons absorbed within the diffusion length to the depletion region and neglect the recombination of the generated electron–hole pairs, the photocurrent I is given by

$$i = e\eta \frac{I}{h\nu} \tag{7}$$

In a p-n junction diode, most of the applied bias voltage appears across the depletion region. Thus, only pairs formed within the region or capable of diffusion into this region can be influenced by the externally applied field and contribute to the external current.

Figure 2 shows a schematic of a simple p-n junction diode. A heavily doped p-type material and a lightly doped n-type material form the p-n junction. Because of different doping levels, the depletion region extends more deeply into the n-type material. At the bottom of the diode is usually a region of n^+-type material which is the substrate. The substrate terminates to the bottom electrical contact. The top electrical contact is fused to the p-type semicon-

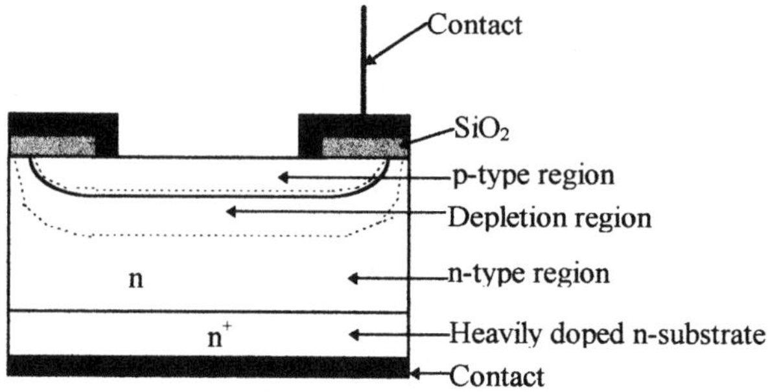

Figure 2 Schematic of a p-n junction dioide.

ductor. An insulating layer of silicon dioxide is provided as shown in Figure 2.

The energy gap of most semiconductors is in the order of 1 eV, which corresponds to an optical wavelength of about 1 μm. Thus, the photodiodes can respond to light in the spectral range from ultraviolet to the near infrared. However, the penetration of the photons through the various regions depends on the frequency of the incident radiation. Ultraviolet radiation is strongly absorbed at the surface, while infrared radiation can penetrate deep into the structure. In a Schottky photodiode, the p-n junction is replaced by a metal semiconductor junction. A thin layer ($<$ 10 nm) of gold is deposited on the n-type semiconductor. This layer is almost trans-

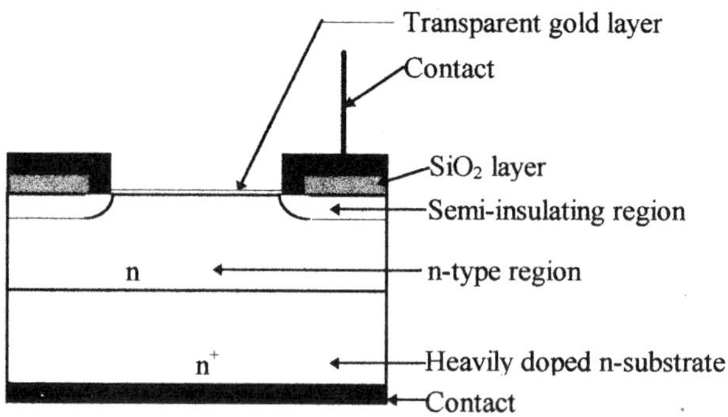

Figure 3 Schematic of a Schottky diode.

parent and forms the metal semiconductor junction. Figure 3 is a schematic of a Schottky diode. Since the depletion region commences right at the semiconductor surface, Schottky photodiodes have superior UV response.

Since the minority carriers may have to travel some distance—up to the diffusion length before being transported across the p-n junction—the response of a p-n junction diode is relatively slow. This is particularly a serious drawback in semiconducting materials such as silicon, in which the depletion region is small compared to the diffusion length. This drawback is overcome by utilizing the p-i-n structure. The thickness of the depletion region can be controlled by doping. In a p-i-n diode, an intrinsic region that has a very high resistivity is sandwiched between the p and n regions. Figure 4a shows the microstructure of a p-i-n diode. To expand the depletion region, the p-i-n diode is usually operated with a reverse bias voltage. Furthermore, the voltage drop occurs mostly across the intrinsic layer (Fig. 4b). When a photon whose energy exceeds the threshold value enters the intrinsic region, electron–hole pairs are created. Under the action of the applied electric field, the photogenerated electrons and holes are swiftly swept toward n- and p-type regions, respectively, and create a signal current. Figure 4c shows a cross section of a typical p-i-n diode. The avalanche photodiode is a junction diode with an internal gain mechanism.

C. Photomultiplier Tube

The PMT is one of the most sensitive optical detectors: photon flux as low as one photon per second can be detected. Many PMT designs exist. Figure 5 shows a cross section of a commonly used PMT consisting of a cathode, a series of dynodes, and an anode in a squirrel cage. A photon of frequency ν incident on the cathode ejects an electron provided $h\nu > \phi$, where ϕ is the work function of the cathode material. A variety of photocathodes cover the spectral range from 0.1 to 11 μm. Figure 6 shows the responsivity and quantum efficiency of some photocathode materials.

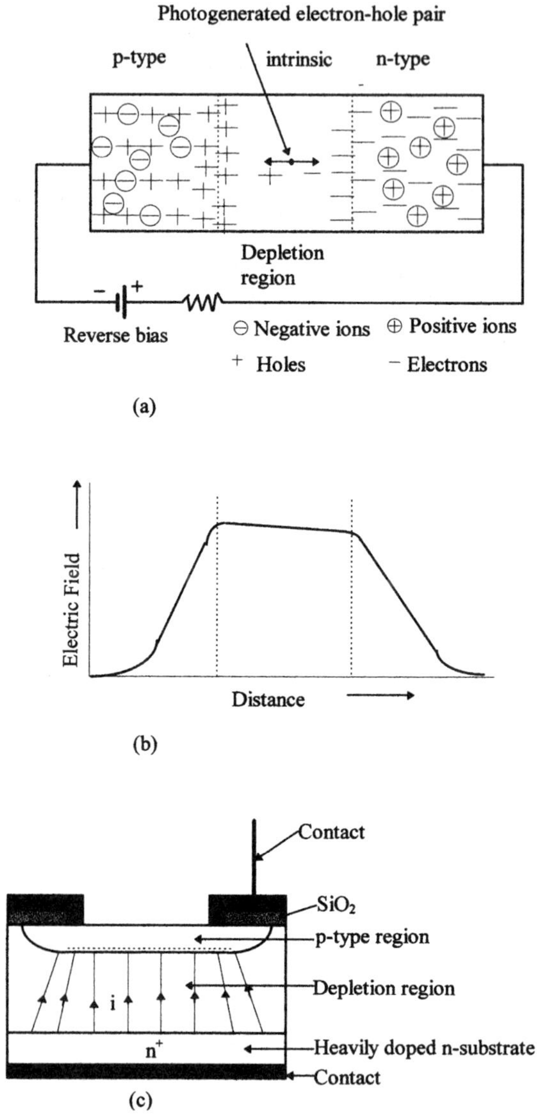

Figure 4 (a) Microstructure of a p-i-n diode. (b) Variation of electric field with distance. (c) Schematic of a p-i-n photodiode.

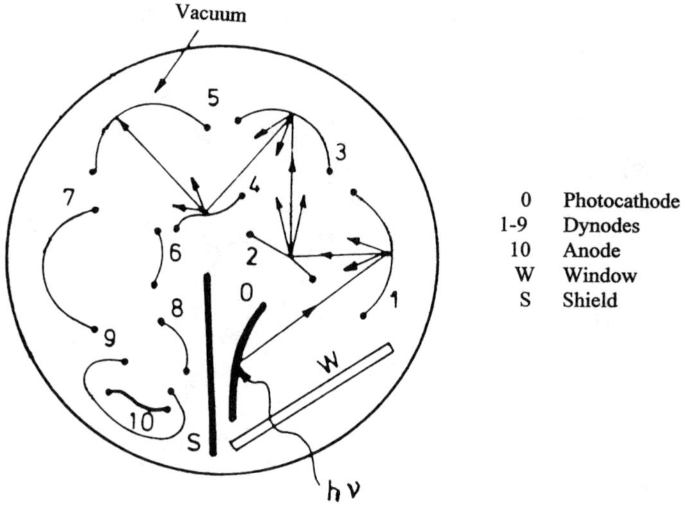

Figure 5 Cross section of a squirrel-cage PMT.

The series of dynodes constitutes a low-noise amplifier. These dynodes are at progressively higher positive potential. A photoelectron emitted from the cathode is accelerated toward the first dynode. Secondary electrons are ejected as a result of the collision of the photoelectron with the dynode surface. This is the first stage of amplification. These secondary electrons are accelerated toward the more positive dynode, and the process is repeated. The amplified electron beam is collected by the anode. The multiplication of electrons at each dynode, or the electron gain A, depends on the dynode material and the potential difference between the dynodes. The total gain G of the PMT is given by

$$G = A^n \tag{8}$$

where n is the number of dynodes. Typical gain is around 1 million. PMTs are extremely fast and sensitive, but they are expensive and require sophisticated associated electronics. These tubes are influenced by stray magnetic fields. However, PMT designs that are not affected by magnetic field exist.

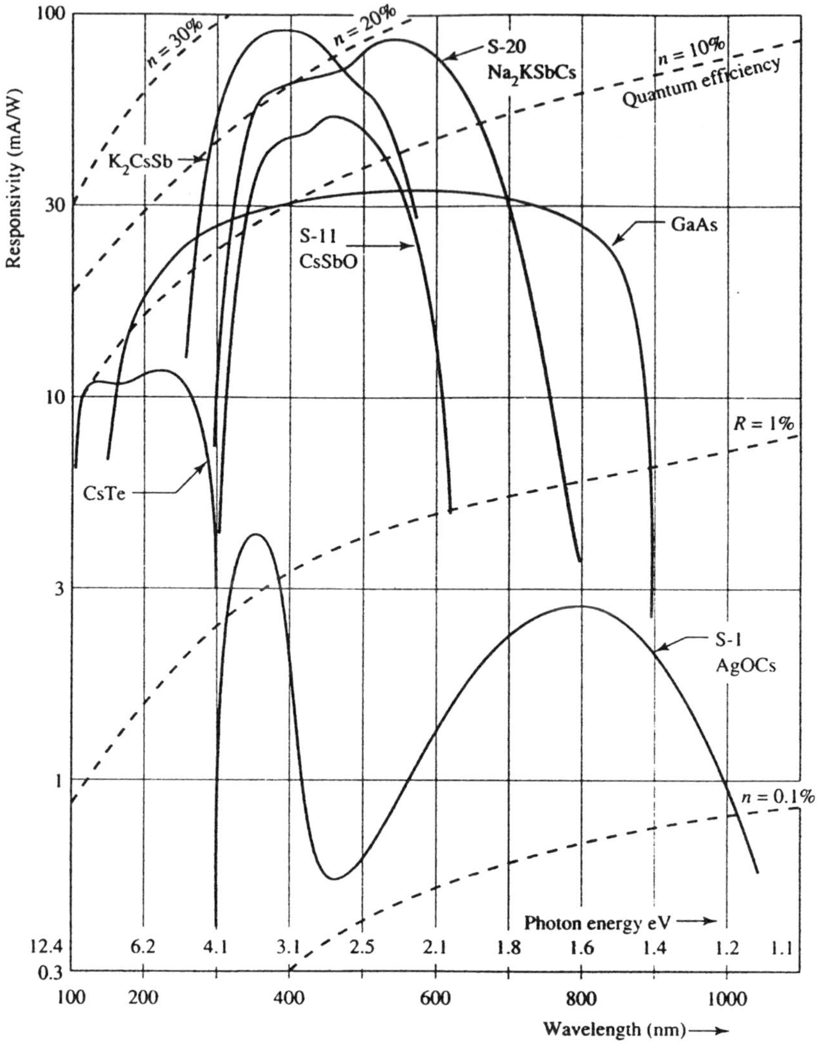

Figure 6 Responsivity and quantum efficiency of some common cathode materials. (From Ref. 8.)

II. IMAGE DETECTORS

The detectors discussed so far are single-element detectors. In many applications [e.g., electronic speckle pattern interferometry (ESPI), robotic vision], the spatially varying information—say, an image—is to be recorded. This is achieved by area array detectors. We can also use a linear array detector to record an image by means of scanning. There are three mechanisms at play in imaging with detector arrays. First, the image is intercepted by the elements of the array and converted to electrical charge distribution, which is proportional to the intensity in the image. Second, the charges are read out as elements of an image while retaining correlation with the position on the array at which each charge was generated. Finally, the image information is displayed or stored.

The area or focal plane arrays can be organized from a single detector, but this implementation requires many wires and processing electronics. The concept of charge-coupled devices (CCDs) makes the retrieval of detector signals easy and eliminates the need for a maze of wires. A charge transfer device in the simplest form is a closely spaced array of metal–insulator semiconductor (MIS) capacitors. The most important, the metal–oxide semiconductor (MOS) capacitor, consisting of silicon and silicon dioxide as the insulator, can be made monolithic.

The basic structure of a CCD is a shift register formed by an array of closely spaced potential well capacitors. A potential well capacitor is schematically shown in Figure 7a. A thin layer of silicon dioxide is grown on a silicon substrate. A transparent electrode is then deposited over the silicon dioxide as a gate, to form a tiny capacitor. When a positive potential is applied to the electrode, a depletion region or an electrical potential is created in the silicon substrate directly beneath the gate. Electron–hole pairs are generated on absorption of incident light. The free electrons generated in the vicinity of the capacitor are stored and integrated in the potential well. The number of electrons (charge) in the well is a measure of the incident light intensity. In a CCD, the charge generated by the incident photons is passed between spatial locations and is detected at

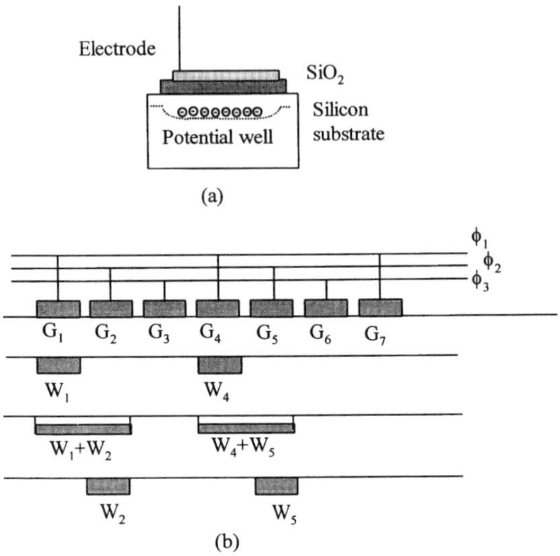

Figure 7 (a) Structure of a CCD pixel. (b) Mechanism of charge transfer in a three-phase clocking layout.

the edge of the CCD. The charge position in the MOS array of capacitors is electrostatically controlled by voltage levels. Under proper application of these voltage levels and their relative phases, the capacitor can be used to store and transfer the charge packet across the semiconductor substrate in a controlled manner.

Figure 7b illustrates the principle of charge transfer through the propagation of potential wells under three-phase clocking layout. In phase ϕ_1, gates G_1 and G_4 are turned on, while all other gates are turned off. Hence, electrons are collected in wells W_1 and W_4. In phase ϕ_2, gates G_1, G_2, G_4, and G_5 are turned on. Therefore, the wells W_1 and W_2, and W_4 and W_5 merge into wider wells. In phase ϕ_3, gates G_1 and G_4 are turned off while G_2 and G_5 are left on. The electrons earlier stored in W_1 are now shifted to W_2. Similarly, the electrons stored in W_4 are now shifted to W_5. This process is repeated until all charge packets have been transferred to the edge of the CCD, where they can be ready by external electronics. Area detector arrays with pixels in numbers of 256 × 256 to 4096 × 4096 are available. The center-to-center distances of the pixels range from 10 to 40 μm, but most commonly used devices have a pixel separation of 12.7 μm. A dynamic range of 1000:1 is quite common. The sensitivity of the video rate CCD cameras is of the order of 10^{-8} W/cm^2, which corresponds to about 0.05 lux. CCD cameras are available with exposure times as small as 1/10,000 second, achieved by electronic shuttering, and SNR of 50 dB.

A. Time-Delay and Integration (TDI) Mode of Operation

Conventional CCD cameras are restricted to working with stationary objects. Object motion during exposure blurs the image. TDI cameras represent a special mode of CCD equipment that provides a solution to the blurring problem. In TDI mode, the charges collected from each row of detectors are shifted to the neighboring sites at a fixed time interval. For example, consider four detectors operating in TDI mode as shown in Figure 8. The object is in motion, and at time t_1, its image is formed on the first detector that creates a charge packet. At time t_2, the image moves to the second detector. Simultaneously, the pixel clock moves the charge packet to the well under the second detector. Here, the image creates additional charge, which is added to the charge shifted from the first detector. Similarly, at time t_3, the image moves to the third detector and creates a charge. Simultaneously, charge from the second detector is moved to the well of the third detector. The charge thus increases linearly with the number N of the detectors in TDI. Therefore, the signal increases with N. The noise also increases but as $\sqrt{N}$, hence signal-to-noise ratio increases as $\sqrt{N}$. The well capacity limits the number of TDI elements that can be used. Clearly, for the camera to work in TDI mode, the charge packet must always be synchronous with the image. Mismatch between image scan rate and clock rate can adversely smear the output.

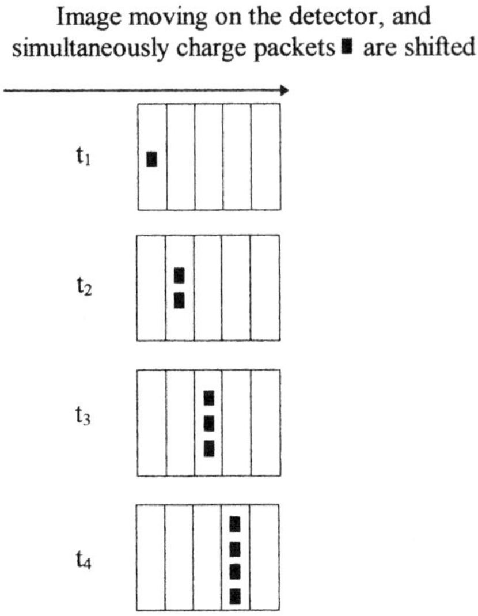

Figure 8 TDI mode of operation.

III. RECORDING MATERIALS

We now discuss the materials that record the image (i.e., the intensity distribution). The recording is required either to keep a permanent record or to provide an input for the measurement and processing. Photographic emulsions are by far the most sensitive and widely used recording media both for photography and holography. Several other recording media have been developed; these are listed in Table 1. We will discuss some features of these media.

A. Photographic Films and Plates

A photographic film or plate consists of silver halide grains distributed uniformly in a gelatin matrix deposited in a thin layer on a transparent substrate (either a glass plate or an acetate film). When the photographic emulsion is exposed to light, the silver halide grains absorb optical energy and undergo a complex physical change: that is, a latent image is formed. The exposed film is then developed. The development converts the halide grains that have absorbed sufficient optical energy into metallic silver. The film is then fixed, which removes the unexposed silver halide grains while leaving the metallic silver. The silver grains are largely opaque at the optical frequency. Therefore, the processed film exhibits spatial variations in opacity depending on the density of silver grains in each region of the transparency.

Photographic emulsions (silver halides) are widely used for photography (incoherent recording) and holography (coherent recording). They are sensitive over a wide spectral range and also offer a very wide resolution range and good dynamic response. An apparent drawback of the photographic materials is that they require wet development and fixing processing, followed by drying. However, development provides a gain of the order of a million, which amplifies the latent image formed during the exposure.

It is mentioned that the spatial variation of light intensity incident on a photoemulsion is converted into the variation of density of metallic silver grains; consequently, its transmission becomes spatially variable. The transmission function τ is related to the photographic density D, the density of the metallic silver grains per unit area, by the relation

$$D = -\log \tau \tag{9}$$

The transmission function τ is defined as

$$\tau(x, y) = \frac{I_t(x, y)}{I_i(x, y)} \tag{10}$$

where $I_t(x, y)$ and $I_i(x, y)$ are the transmitted and incident intensities, respectively. The reflection losses at the interfaces are ignored. The transmission is averaged over a very tiny area around the point (x, y).

One of the most commonly used descriptions of the photosensitivity of a photographic film is the Hurter–Driffield curve or H&D curve. It is the plot of the density D versus the logarithm of the exposure E. The exposure E is defined as the energy per unit area

Table 1 Recording Media for Photography and Holography

Class of material	Spectral range (nm)	Recording process	Spatial frequencies (l/mm)	Types of grating	Processing	Readout process	Maximum diffraction efficiency (%)
Photographic materials	400–700 (< 1300)	Reduction to Ag metal grain;	> 7000	Plane/volume amplitude	Wet chemical	Density change	5
		bleached to silver salts		Plane/volume phase	Wet chemical	Refractive index change	20–50
Dichromated gelatin	250–520 and 633	Photo-cross-linking	> 3000	Plane phase	Wet chemical followed by heat	Refractive index change	30
				Volume phase			> 90
Photoresists	UV–500	Photo-cross-linking or photo-polymerization	< 3000	Surface relief/phase blazed reflection	Wet chemical	Surface relief	30–90

Photo-polymers	UV–500	Photo-polymerization	~200–1500 bandpass	Volume phase	None or post exposure and postheating	Refractive index change or surface relief	10–85
Photoplastics/ photo-conductor thermo-plastics	Nearly pan-chrom-atic for PVK TNK photo-conduc-tor	Formation of an electrostatic latent image with electric-field-produced deformation of heated plastic	400–1000 band pass	Plane phase	Corona charge and heat	Surface relief	6–15
Photo-chromics	300–450	Generally photo-induced new absorption bands	> 2000	Volume absorption	None	Density change	1–2
Ferroelectric crystals	400–650	Electrooptic effect/ photo-refractive effect	> 3000	Volume phase	None	Refractive index change	90

incident on the film and expressed as $E = IT$, where I is the incident intensity and T is the exposure time. Figure 9 illustrates a typical H&D curve for a photographic negative. If the exposure is below a certain level, the density is independent of the exposure. This minimum density is usually referred to as gross fog. In the "toe," the density begins to increase with log E. There follows a region in which the density increases linearly with logarithm exposure; this is the linear region of the curve and the slope of this linear region is referred to as the film gamma (γ). Finally, the curve saturates in a region called the shoulder. There is no change in density with logarithm exposure. However, with very large exposures, solarization takes place.

The linear region of the H&D curve is generally used in conventional photography. A film with a large value of γ is called a high contrast film, while a film with low γ is a low contrast film. The gamma of the film also depends on development time and the developer. It is thus possible to obtain a prescribed value of γ by the judicious choice of film, developer, and development time.

Since the emulsion is usually used in the linear portion of the H&D curve, the density D when the film is given an exposure E can be expressed as follows:

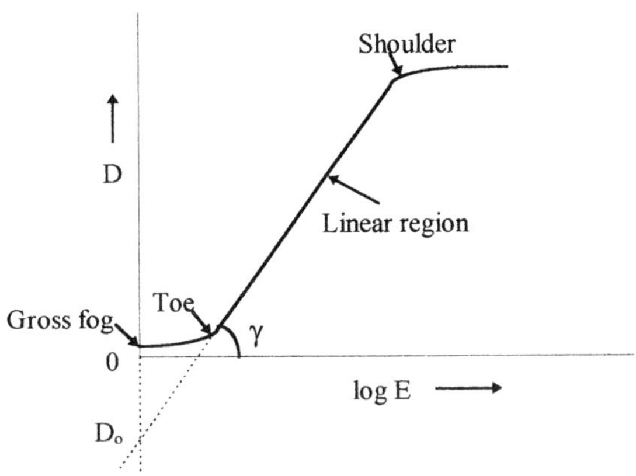

Figure 9 The Hurter–Driffield (H&D) curve.

$$D = \gamma_n \log E - D_0 = \gamma_n \log (IT) - D_0 \tag{11}$$

where the subscript n means that a negative film is being used and D_0 is the intercept. Equation (11) can be written in terms of the transmission function τ_n of the negative film as follows:

$$\tau_n = K_n I^{-\gamma_n} \tag{12}$$

where $K_n = 10^{D_0} T^{-\gamma_n}$ is a positive constant. Equation (12) relates the incident intensity to the transmission function of the film after development. It is seen that the transmission is a highly nonlinear function of the incident intensity for any positive value of γ_n. In many applications, it is required either to have a linear or a power law relationship. This, however, requires a two-step process. In the first step, a negative transparency is made in the usual fashion, which will have a gamma γ_n. The transmission of this negative transparency is given by

$$\tau_{n1} = K_{n1} I^{-\gamma_{n1}} \tag{13}$$

In the second step, the negative transparency is illuminated by uniform intensity I_0 and the light transmitted is used to expose a second film. This results in a positive transparency with transmission τ_p, where

$$\tau_p = K_{n2}(I_0 \tau_{n1})^{-\gamma_{n2}} = K_{n2} I_0^{-\gamma_{n2}} K_{n1}^{-\gamma_{n2}} I^{\gamma_{n1}\gamma_{n2}} = K_p I^{\gamma_p} \tag{14}$$

where K_p is another positive constant and $\gamma_p = \gamma_{n1}\gamma_{n2}$ is the overall gamma of the two-step process. Evidently, a positive transparency does provide a linear mapping of intensity when the overall gamma is unity.

When photographic emulsions are used for holography or in general, for coherent optical systems, the H&D curve is never used. Instead, the plot of amplitude transmittance $\mathbf{t}(x, y)$ versus exposure E is used. The amplitude transmittance is usually complex, since film introduces both amplitude and phase variations to the incident plane wave. However, if the film is used in a liquid gate, the phase variations can be eliminated. The amplitude transmittance then is given by the square root of the transmission function: that is, $|\mathbf{t}(x, y)| = \sqrt[+]{\tau(x, y)}$. Typical plots of $|\mathbf{t}(x, y)|$ versus E for several holographic emulsions are shown in Figure 10. Holographic recording is generally carried out in the linear region of the $|\mathbf{t}(x, y)|$ versus E curve.

Since formation of the latent image does not cause any changes in the optical properties during exposure, it is possible to record several holograms in the same photographic emulsion without any interaction between them. The information can be recorded in the form of either amplitude transmittance or phase variations. For recording in the form of phase variations, the amplitude holograms are bleached. Bleaching converts the metallic silver grains back to the transparent silver halide crystals. Also, holographic infomation can be recorded in the volume, provided the emulsion is sufficiently thick and has enough resolution.

B. Dichromated Gelatin

Dichromated gelatin is, in some respects, an ideal recording material for volume phase holograms: it has large refractive index modulation capability, high resolution, and low absorption

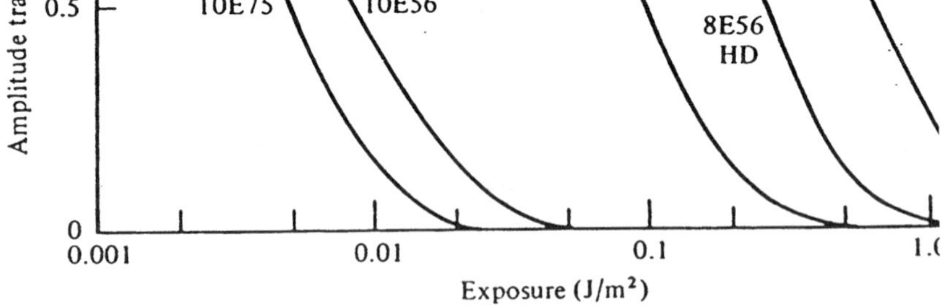

Figure 10 Plots of amplitude transmittance $|\mathbf{t}|$ versus exposure E for several holographic emulsions.

and scattering. The gelatin layer can be deposited on a glass plate and sensitized. Alternately, the photographic plates can be fixed, rinsed, and sensitized. Ammonium, sodium, and potassium chromates have been used as sensitizers. Most often, ammonium dichromate is used for sensitization. Dichromated gelatin exhibits sensitivity in the wavelength range of 250–580 nm. The sensitivity at 514 nm is about a fifth of that at 448 nm. Gelatin can be sensitized at the red wavelength of a He–Ne laser by the addition of methyl blue dye.

The exposure causes cross-linking between gelatin chains and alters swelling properties and solubility. Treatment with warm water dissolves the unexposed gelatin and thereby forms a surface relief pattern. However, much better holograms can be obtained if the gelatin film is processed to obtain a modulation of the refractive index. During processing, rapid dehydration of the gelatin film is carried out in an isopropanol bath at elevated temperature. This creates a very large number of small vacuoles in the gelatin layer, hence modulates the refractive index. It is also suggested that the formation of complexes involving Cr^{3+} compound, gelatin, and isopropanol in the hardened areas is also partly responsible for the refractive index modulation. The phase holograms in gelatin are very efficient, directing more than 90% of the incident light into the useful image.

C. Photoresists

Photoresists are organic materials that are sensitive in the ultraviolet and blue regions. They are all relatively slow and thus require very long exposure. Usually, a thin layer ($\approx 1\ \mu m$) is obtained either by spin-coating or spray-coating on the substrate. This layer is then baked at around 75°C. On exposure, one of three processes takes place: formation of an organic acid, photo-cross-linking, or photopolymerization.

There are two types of photoresist—negative and positive. In negative photoresists, the unexposed regions are removed during development, while in positive photoresists, the exposed regions are removed during development. Therefore, a surface relief recording is obtained. A grating recorded on photoresist can be blazed by ion bombardment. Blazed grating can also be recorded by optical means. The relief recording offers the advantage of replication using thermoplastics.

When exposure is made in a negative photoresist from the air–film side, the layer close to the substrate is the last to photolyze. This layer will be simply dissolved away during development, since it has not photolyzed fully. This nonadhesion of negative photoresists in holographic recording is a serious problem, which is overcome by exposing the photoresist from the substrate–film side to obtain better photolysis of the resist at the substrate–film interface. On the other hand, the positive photoresists do not have this problem, hence are preferred. One of the most widely used positive photoresists is Shipley AZ-1350. The recording can be done at 458 nm of Ar^+ laser or at 442 nm wavelength of He–Cd laser.

D. Photopolymers

Photopolymers are also organic materials. Photopolymers for use in holography can be in the form of either a liquid or a dry layer enclosed between glass plates. Exposure causes photopolymerization or cross-linking of the monomer, resulting in refractive index modulation that may or may not be accompanied by surface relief. Photopolymers are more sensitive than photoresists, hence require moderate exposure. They also possess the advantage of dry and rapid processing. Thick polymer materials like PMMA (polymethyl methacrylate) and CAB are excellent candidates for volume holography because the refractive index change on exposure could be as large as 10^{-3}. Two photopolymers are commercially available; these are Polaroid DMP 128 and du Pont OmniDex. Polaroid DMP 128 uses dye-sensitized photopolymerization of a vinyl monomer incorporated in a polymer matrix, which is coated on a glass or plastic substrate. Coated plates or films can be exposed with blue, green, and red light. du Pont OmniDex film, which consists of a polyester base coated with a photopolymer, is used for contact copying of master holograms by means of UV radiation.

The response of photopolymers is band-limited because of the limitation imposed by the diffusion length of the monomer at the lower end and the length of the polymer at the higher end of the response curve.

E. Thermoplastics

A thermoplastic recording material is a multilayer structure having a substrate coated with a conducting layer, a photoconductor layer, and a thermoplastic layer. A photoconductor that works well is the polymer poly-*N*-vinyl carbazole (PVK) to which is added a small amount of electron donor 2,4,7-trinitro-9-fluorenone (TNF). The thermoplastic layer is of a natural tree resin, staybelite. The thermoplastics for holographic recording combine the advantages of high sensitivity and resolution, dry and nearly instantaneous in situ development, erasability, and high readout efficiency.

The recording process involves a number of steps. First, a uniform electrostatic charge is established on the surface of the thermoplastic in a darkened chamber by means of a corona discharge assembly. The charge is capacitively divided between the photoconductor and the thermoplastic layers. In the second step, the thermoplastic is exposed; the exposure causes the photoconductor to discharge its voltage at the illuminated regions. This does not cause any variation in the charge distribution on the thermoplastic layer; the electric field in the thermoplastic layer remains unchanged. In the third step, the surface is charged again by the corona discharge assembly. In this process, the charge is added at the exposed regions. Thus one now has an electric field distribution, which forms a latent image. In the fourth step, the thermoplastic is heated to its softening point, thereby developing the latent image. The thermoplastic layer undergoes local deformation as a result of the varying electric field across it, becoming thinner wherever the field is higher (the illuminated regions) and thicker in the unexposed areas. Cooling quickly to room temperature freezes the deformation; the recording is now in the form of surface relief. The recording is stable at room temperature but can be erased by heating the thermoplastic to a temperature higher than that used for development. At the elevated temperature, surface tension evens out the thickness variation, hence erases the recording. This is the fifth step; the erasure step. Figure 11 shows the whole

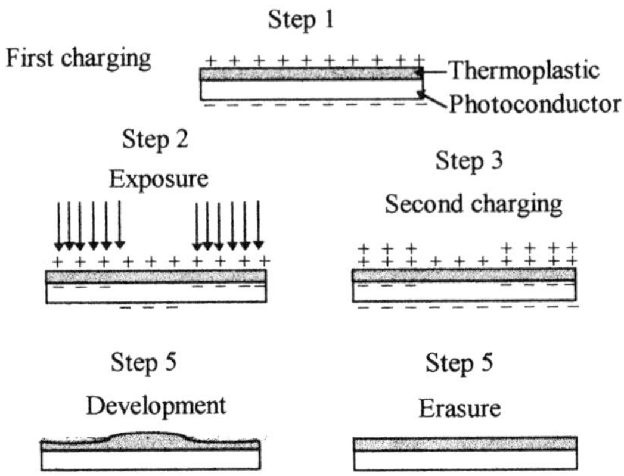

Figure 11 The record–erase cycle for a thermoplastic material.

recording process. The thermoplastic can be reused several hundred times. The response of these devices is band-limited, depending on the thickness of the thermoplastic and other factors.

F. Photochromics

Materials that undergo a reversible color change upon exposure are called photochromics. Photochromism occurs in a variety of materials, both organic and inorganic. Organic photochromics have a limited life and are prone to fatigue. However, organic films of spiropyrane derivatives have been used for hologram recording in darkening mode at 633 nm. Inorganic photochromics are either crystals or glasses doped with selected impurities; photochromism is due to reversible charge transfer between two species of electron traps. Recording in silver halide photochromic glasses has been done in darkening mode at 488 nm and in bleaching mode at 633 nm. Doped crystals of CaF_2 and SrO_2 have been used in bleaching mode at 633 nm.

The sensitivity of photochromics is very low because the chemical reaction occurs at the molecular level. For the same reason, they are essentially grain-free and have resolution in excess of 3000 l/mm. Inorganic photochromics have large thicknesses, and thus a number of holograms can be recorded in them. They do not require any processing and can be reused almost indefinitely. In spite of all these advantages, these materials have limited applications because of their low diffraction efficiency (< 0.02) and low sensitivity.

G. Ferroelectric Crystals

Certain ferroelectric crystals [e.g., lithium niobate ($LiNbO_3$), lithium tantalate ($LiTaO_3$), barium titanate ($BaTiO_3$), strontium barium niobate (SBN)], exhibit small changes in refractive index when exposed to intense light. The photoinduced refractive index change can be reversed by an application of heat and light. The mechanism of recording in these crystals is as follows: exposure to light frees trapped electrons, which then migrate through the crystal lattice and are again trapped in adjacent unexposed or low intensity regions. The migration usually occurs through diffusion or an internal photovoltaic effect. This produces a spatially varying net space–charge distribution and corresponding electric field distribution. The electric field modulates the refractive index through the electrooptic effect and creates a volume phase grating. These are real-time recording materials; the records are stable as the charges are bound to the localized traps. The recording can, however, be erased by illuminating it with a light beam of wavelength that can release the trapped electrons.

Lithium niobate crystal (particularly Fe-doped) has been used for holographic interferometry and data storage. Recording is fixed by temperature. This crystal is rather slow, however. Higher sensitivity is obtained with photoconductive electrooptic crystals like BSO and BGO by an application of an external electric field. These crystals are available in the form of thin slices several centimeters in diameter. Barium titanate crystal is used in holographic interferometry and speckle photography because of its very slow response. Recording can be made over a very wide spectral range.

BIBLIOGRAPHY

1. Joseph W. Goodman, *Introduction to Fourier Optics*, McGraw-Hill, New York (1968, 1996).
2. H. M. Smith, *Holographic Recording Materials*, Springer-Verlag, Berlin (1977).
3. Eustace L. Dereniak and Devon G. Crowe, *Optical Radiation Detectors*, John Wiley, New York (1984).
4. Rajpal S. Sirohi and Mahendra P. Kothiyal, *Optical Components, Systems, and Measurement Techniques*, Marcel Dekker, New York (1991).
5. M. P. Petrov, S. I. Stepanov, and A. V. Khomenko, *Photorefractive Crystals in Coherent Optical Systems*, Springer-Verlag, Berlin (1991).
6. James T. Luxon and David E. Parker, *Industrial Lasers and Their Applications*, Prentice Hall, Englewood Cliffs, NJ (1992).
7. H. I. Bjelkhagen, *Silver-Halide Recording Materials for Holography and Their Processing*, Springer-Verlag, Berlin (1993).
8. Endel Uiga, *Optoelectronics*, Prentice Hall, Englewood Cliffs, NJ (1995).
9. Sudhanshu S. Jha (Ed.), *Perspectives in Optoelectronics*, World Scientific Publishing, Singapore (1995).
10. James E. Stewart, *Optical Principles and Technology for Engineers*, Marcel Dekker, New York (1996).
11. P. Hariharan, *Optical Holography*, Cambridge University Press, Cambridge (1996).
12. Hans I. Bjelkhagen (Ed.), *Holographic Recording Materials*, Society of Photo-Optical Instrumentation Engineers Milestone MS 130, SPIE Optical Engineering Press, Bellingham, WA. (1996).
13. Gerald C. Holst, *CCD Arrays, Cameras, and Displays*, JCD Publishing and SPIE Optical Engineering Press, Bellingham, WA. (1996).
14. Francis T. S. Yu and Xiangyang Yang, *Introduction to Optical Engineering*, Cambridge University Press, Cambridge (1997).

Chapter 5

Holographic Interferometry

I. INTRODUCTION

Holography was born out of a very challenging technological problem: the need to improve the resolution of the electron microscope, which was limited by the spherical aberration of the electron lenses. Gabor, therefore, invented a two-step process, the first step involving recording without the lenses and the second step as reconstruction. The technique was demonstrated with micro-objects and with spatially and temporally filtered radiation from a mercury lamp. This arrangement was necessary because high resolution recording materials and coherent sources were not available at that time.

After its invention in 1948, holography remained practically dormant until the arrival of the laser, since a long coherence length source was needed to record a hologram of an object. Earlier recordings of three-dimensional objects were made on Kodak 649F plates, and the results were very impressive. Holography, therefore, came to be known as three-dimensional photography.

It, however, is more than ordinary 3D photography, since it provides 3D views with changing perspectives.

Holography records the complex amplitude of a wave coming from an object rather than the intensity distribution in the image, as is the case in photography. Holography literally means total recording, that is, recording of both the phase and the amplitude of a wave. The recording of the complex amplitude is accomplished by interferometry inasmuch as all the detectors in the optical regime are energy detectors. Therefore, a reference wave is added to the object wave at the recording plane. The recording is done on a variety of media (photoemulsions, photopolymers, thermoplastics, etc.). The record, called a hologram, is like a window with a memory. Different perspectives of the scene are seen through different portions of the hologram. In addition to recording holograms of 3D objects, several new applications of holography have emerged, holographic interferometry being one of these.

Holographic interferometry (HI), which provides interferometric comparison of real objects or events separated in time and space, is a technique of unparalleled applications. Various kinds of HI have been developed—for example, real-time, double-exposure, and time-averaged HI. Further, it can be performed with one, two, or more reference waves, which can be of the same or different wavelengths. The reference wave can come from the same side of the hologram as the object wave or from the other side of the hologram. HI can be performed with a continuous wave laser or a pulsed laser. The record can be made on photoemulsions, thermoplastics, photopolymers, and other media. The possibilities are endless, and so are the applications. This chapter presents some of these applications along with the relevant theoretical background.

II. HOLOGRAM RECORDING

An object is illuminated by a wave from a laser and the diffracted field is received on a recording plane lying in the x–y plane. A reference wave is added to this field at the recording plane as shown in Figure 1a. The diffracted field from the object constitutes

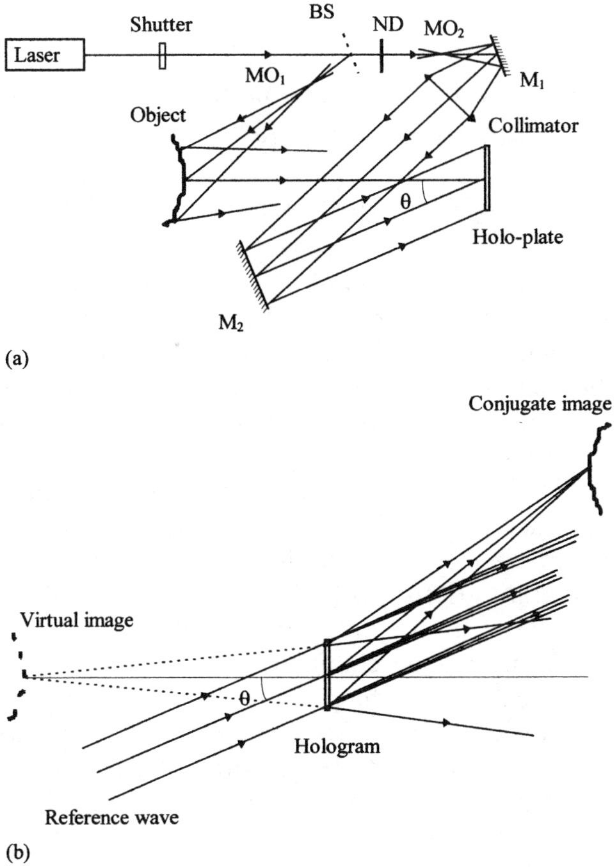

Figure 1 (a) Recording of a hologram and (b) its reconstruction.

the object wave, which is represented by $\mathbf{O}(x, y) = O_o(x, y) \exp i\phi_o(x, y)$, where $O_o(x, y)$ is the amplitude of the object wave and $\phi_o(x, y)$ its phase. The complex amplitude $\mathbf{R}(x, y)$ of the reference wave at the recording plane is expressed as $\mathbf{R}(x, y) = R \exp 2\pi i \nu_R y$, where ν_R is the spatial frequency of the wave. We have taken a plane wave incident at an angle of θ with the z axis as a reference wave, hence $\nu_R = (\sin\theta)/\lambda$. These waves are derived from the same wave (source), hence are coherent to each other. The total amplitude at the recording plane is

$$\mathbf{A}(x, y) = \mathbf{O}(x, y) + \mathbf{R}(x, y) \tag{1}$$

Therefore, the intensity distribution $I(x, y)$ on the recording plane is given by

$$I(x, y) = O_o^2(x, y) + R^2 + 2O_o(x, y)R\cos\{2\pi\nu_R y - \phi_o(x, y)\} \tag{2}$$

It is thus seen that both the amplitude $O_o(x, y)$ and the phase $\phi_o(x, y)$ variations have been converted into intensity variations to which the recording material responds. We assume that the recording material is a photoemulsion, say, a holographic plate or a film. The intensity is recorded over an appropriate time period T resulting in an exposure variation $E(x, y) = I(x, y)T$. After development, the plate or film is called a hologram. On the hologram, the exposure variation is converted into density variation or amplitude transmittance variation. The amplitude transmittance is complex because the hologram introduces both the amplitude and the phase variations; the phase variation as a result of thickness variation. The phase variation can be eliminated if the hologram is placed in a liquid gate. Further, if the phase variations are shared between two wave fronts, they are not seen when these wave fronts interfere. We, however, assume that the amplitude transmittance of the hologram is proportional to the exposure incident during recording. Under this assumption, the amplitude transmittance $t(x, y)$ of the hologram is expressed as follows:

$$t(x, y) = t_0 - \beta E(x, y) \tag{3}$$

where β is a constant dependent on processing parameters, exposure, and so on.

III. RECONSTRUCTION

The hologram so recorded is placed back in the same position held during recording and illuminated by the reference wave. The field just behind the hologram is given by

$$t(x, y)\mathbf{R}(x, y) = t_0'\mathbf{R}(x, y) - \beta T\mathbf{R}(x, y)[O_o^2(x, y) + 2O_o(x, y)R\{\cos 2\pi\nu_R y - \phi_o(x, y)\}] \tag{4}$$

where $t_0' = t_0 - \beta TR^2$ is the modified dc transmittance. Equation (4) can also be written as follows:

$$t(x, y)\mathbf{R}(x, y) = t_0'\mathbf{R}(x, y) - \beta T\mathbf{R}(x, y)[O_o^2(x, y) + \mathbf{O}(x, y)\mathbf{R}^*(x, y) + \mathbf{O}^*(x, y)\mathbf{R}(x, y)] \tag{5}$$

where an asterisk signifies a complex conjugate. It is seen that there are four waves just behind the hologram of which the wave $t_0'\mathbf{R}(x, y)$ is the reference wave uniformly attenuated. The wave $-\beta T\mathbf{R}(x, y)O_o^2(x, y)$ also propagates in the direction of the reference wave. Because $O_o^2(x, y)$ varies slowly in space, there is a diffracted field around the direction of the reference wave. The third wave $-\beta TR^2\mathbf{O}(x, y)$ is the original object wave multiplied by a constant, $-\beta TR^2$. This wave propagates in the direction of the object wave and has all the attributes of the object wave except that its spatial dimensions are restricted by the size of the hologram. The negative sign signifies that a phase change of π has taken place because of the wet development process. The fourth wave $-\beta TR^2(x, y)\mathbf{O}^*(x, y)$ represents a conjugate wave that propagates in a different direction. This wave can also be written as $-\beta TR^2O(x, y)\exp[2\pi i 2\nu_R y - \phi_o(x, y)]$; the phase of the reference wave is modulated by that of the object wave, but it essentially travels in the direction ϕ, where $\sin\phi = 2\sin\theta$. Figure 1b shows that several waves are generated during the reconstruction step.

The recording and reconstruction geometries just described are the off-axis holography geometries due to Leith and Upatnieks. However, the reference beam can be added axially to the object beam. This is in-line holography or Gabor holography. In-line holography has applications in, for example, particle size measurements.

IV. CHOICE OF ANGLE OF REFERENCE WAVE

One of the shortcomings of Gabor holography is that all the diffracted waves propagate in the same direction. Their separation was effected by adding the reference wave at an angle. The question now is, How large should the angle be? In fact, the off-axis reference wave acts as a carrier of object information. It also results in very fine interference fringes into which object information is coded. The fringe frequency is given by $\sin\theta/\lambda$, where θ is the mean angle between the reference and object waves. The recording medium should be capable of resolving this fringe frequency. Further, as described earlier, the various waves are angularly separated by nearly θ for small angles. But a simple condition that forbids overlapping of the spectra of the various waves restricts θ. If the bandwidth of the object wave is ν_o then it is easily seen from Figure 2 that the spectra will be just separated when $3\nu_o = \nu_R$. This immediately gives the smallest angle θ_{min} as $\theta_{min} = \sin^{-1}(3\lambda\nu_o)$. For the angles smaller than θ_{min}, the various diffracted beams will overlap.

V. CHOICE OF REFERENCE WAVE INTENSITY

It was tacitly assumed that the amplitude transmittance of the hologram is proportional to the exposure. This assumption is valid only when one operates over a very small portion of the *t*–*E* curve shown in Figure 3. To meet this condition, the reference wave should be 3 to 10 times stronger than the object wave. If this

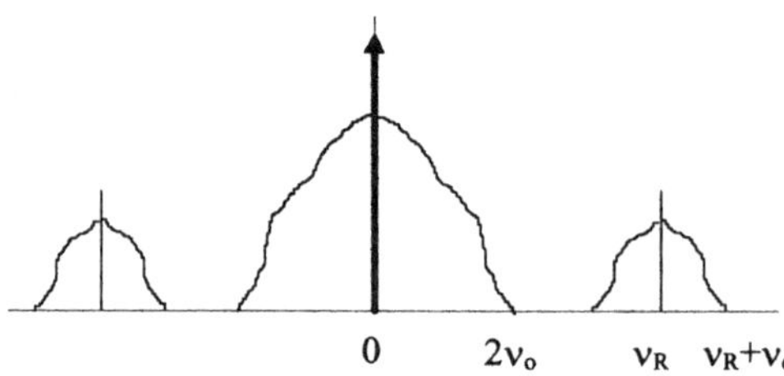

Figure 2 Fourier transform of the transmittance of a hologram.

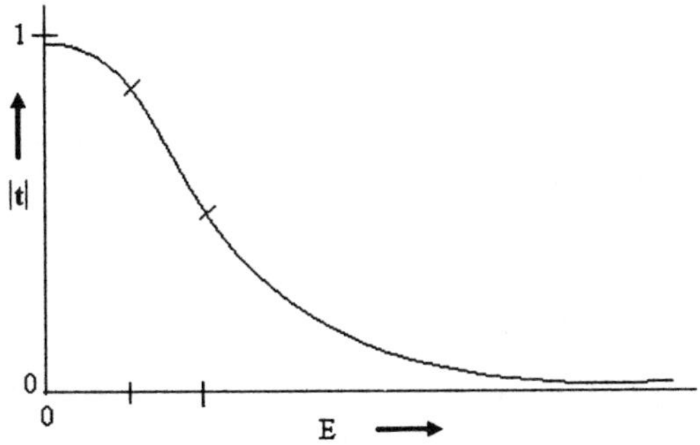

Figure 3 Amplitude transmittance versus exposure (|**t**| vs. E) curve for a typical photoemulsion.

condition is not met, nonlinearities will begin to play a role, and higher order images are produced.

VI. TYPES OF HOLOGRAMS

Holograms are classified by a variety of ways, as summarized in Table 1.

VII. DIFFRACTION EFFICIENCY

Diffraction efficiency is a measure of the amount of light that goes into forming a useful image when the hologram is illuminated by the reference wave. It is calculated for the ideal situation of plane wave interference. For a thin-amplitude transmission hologram, the maximum diffraction efficiency is 6.25%. This value can be improved by bleaching (i.e., the amplitude hologram is converted into a phase hologram). Again, it can reach up to 33.6% for a thin-phase transmission hologram. The diffraction efficiency for a thick-phase hologram can approach 100%: essentially then, all the incident light is utilized for image formation and there is only one diffraction order.

Table 1 Classification of Hologram Types

Properties	Holograms Types
Transmission function	Amplitude hologram
	Phase hologram
Region of diffraction	Fresnel hologram
	Fraunhofer hologram
	Fourier hologram
	Image plane hologram
Recording geometry	On-axis/Gabor/in-line hologram
	Off-axis/Leith–Upatnieks/carrier frequency hologram
	Reflection/Denisyuk hologram
Exposure	Single-exposure/real-time/live-fringe hologram
	Double-exposure/lapsed time/frozen-fringe hologram
	Multiple-exposure hologram
	Time-averaged hologram
Emulsion thickness/ Q-parameter	Thin hologram
	Thick hologram
Reconstruction	Monochromatic light hologram
	White light hologram
Recording geometry	Transmission hologram
	Reflection hologram
	Rainbow hologram
	Digital hologram
Object type	Transmission objects/phase object hologram
	Transmission objects–diffuse illumination hologram
	Opaque objects hologram

VIII. EXPERIMENTAL ARRANGEMENT

The experimental arrangement (Fig. 1a) consists of a laser, a beam splitter (BS), beam-expanding optics, and the recording medium. We describe these in detail.

A. Lasers

There are a large number of lasers to choose from. Some of these are listed briefly.

1. *Ruby laser*: a pulsed laser with a pulse width of the order of 10 ns. Its coherence length may exceed 2 m when used with an etalon, and it can deliver more than 1 joule in a single pulse. The laser for HI has a dual pulse facility with variable pulse separation for dynamical studies.
2. *Argon ion laser*: a continuous wave laser with output exceeding 5 W. It gives a multiline output, hence a dispersion prism is mounted to select a line. Further, an intercavity etalon increases the coherence length to be usable for studying moderate-sized objects. It can be used with photoresist and thermoplastics as recording media.
3. *He–Ne laser*: the most commonly used laser for holography and HI, with power outputs in the range of 25–35 mW. The coherence length is more than 20 cm.
4. *Nd:YAG semiconductor pumped, first harmonic (green)*: a laser that is finding acceptance for holography; it offers a very long coherence length with continuous wave operation. The output ranges from 40 to 150 mW.
5. *He–Cd laser*: a continuous wave laser with output in the range of 40 to 150 mW. It is suitable for recording on photoresist. Both blue and ultraviolet lines are usable for holography.
6. *Semiconductor or diode lasers (LDs)*: can be run both as continuous and as pulsed. Continuous wave lasers are preferred for holography but are not really very common. They require temperature and current stability. They can output up to 500 mW continuous wave.

B. Beam Splitters

Usually an unexpanded laser beam is split into two or more beams. For this, a glass plate (preferably a wedge plate) serves as a good beam splitter. However, when an expanded collimated beam is to be split, a cube beam splitter or a plane parallel plate splitter coated on one side with antireflection layer is recommended.

C. Beam Expanders

With high power lasers, a suitable concave or negative lens is used for beam expansion; otherwise, a microscope objective 5×, 10×, 20×, 40 × (45×), 60 × (63×) could be used. Since dust particles tend to collect on the optical surfaces, the beam is usually dirty (i.e., it has circular rings, specks, etc.). It can be cleaned by placing a pinhole at the focus of the microscope objective. The arrangement is known as spatial filtering. The pinhole size is to be matched with the microscope objective. An achromatic lens is placed in the beam such that the pinhole is at its focus. This results in an expanded collimated beam; hence the name beam expander.

D. Object Illumination Beam

For small-sized objects, a collimated beam illumination is preferred because the propagation vector for the illumination beam is then constant over its whole surface. For large objects, a spherical diverging wave is used; a cylindrical diverging wave is preferred if the object is cylindrical. Objects must be located in the coherence volume. Special geometries that relax the coherence requirement may be used.

E. Reference Beam

Either a diverging spherical wave or a plane wave can be used as a reference wave. If the hologram is to be reconstructed at a later time, or at a different location, however, use of a plane wave is recommended. The beam is overexpanded to ensure uniformity over the hologram plane. The reference beam intensity must be 3 to 10 times stronger than the object beam intensity at the hologram plane for a linear recording.

F. Angle Between Object and Reference Beams

The object wave is a diffuse wave. Therefore, the angle between the object wave and the reference wave (even when it is a plane wave) will vary over the recording plane. We, however, take the mean angle. The fringe spacing on the recording plane will be approximately $\lambda/\sin\theta$, where θ is the mean angle between the reference wave and the object wave incident nearly normally on the recording plane. This fringe pattern should be resolvable by the recording medium. Therefore, too large an angle puts higher demand on the resolution. A small angle may not result in the separation of the beams. Therefore, the angle is to be chosen judiciously.

IX. HOLOGRAPHIC RECORDING MATERIALS

A variety of recording materials are available, as described in Chapter 4. The choice is made based on the wavelength sensitivity, resolution, sensitivity, and other properties.

X. HOLOGRAPHIC INTERFEROMETRY

The comparison of two wave fronts from the real object is accomplished with holographic interferometry. These two waves usually are from the initial (unstressed) state and the final (stressed) state of the object. It is, however, assumed that the microstructure of the surface does not change as a result of loading: the two comparison waves differ because of path difference changes rather than microstructural changes. HI can be performed in a variety of ways. The most common methods are real-time, double-exposure, and time-average HI. Several new configurations have been developed depending on the application and also to exploit the strengths of HI. They are discussed at appropriate places.

A. Real-Time HI

A hologram of the object is made, processed, and placed back exactly at the same location it occupied during recording. Reconstruction of the hologram by the reference wave generates a replica of the original object wave, which propagates in the direction of the original wave. Owing to the wet photographic development, however, this wave is phase-shifted by π. If the object wave is also present, it will be transmitted by the dc transmittance of the hologram: these two waves thus interfere, giving a dark field, since they are identical in all respects but for the phase change of π.

If the object is now loaded, the object wave carries the deformation phase, and one observes an interference pattern, which changes in real time with the load. One can therefore monitor the response of the object to the external loading agency continuously until the fringes become too fine to be resolved. The technique is also known as single-exposure or live-fringe HI.

Mathematically, we can explain the procedure by writing the transmittance of the single-exposure hologram as follows:

$$t(x, y) = t_0' - \beta T[O_o^2(x, y) + 2O_o(x, y)R\cos\{2\pi\nu_R y - \phi_o(x, y)\}] \tag{6}$$

with $t_0' = t_0 - \beta TR^2$. Interrogation by the reference wave $\mathbf{R}(x, y)$ releases the object wave which is now given as $-\beta TR^2\mathbf{O}(x, y)$, while the directly transmitted object wave is written as $\{t_0' - \beta TO_o^2(x, y)\}\mathbf{O}(x, y)e^{i\delta}$, where δ is the phase change introduced by the deformation. Interference between these two waves results in an intensity distribution that is expressed as follows:

$$I = (\beta TR^2)^2 O_o^2\left[1 + \frac{(t_0 - \beta TR^2 - \beta TO_o^2)^2}{(\beta TR^2)^2} - 2\frac{(t_0 - \beta TR^2 - \beta TO_o^2)}{\beta TR^2}\cos\delta\right] \tag{7}$$

Assuming $\beta TR^2 \gg \beta TO_o^2$—which is usually true—Eq. (7) is further simplified as follows:

$$I = I_o\left[1 + \frac{(t_0 - \beta TR^2)^2}{(\beta TR^2)^2} - 2\frac{(t_0 - \beta TR^2)}{\beta TR^2}\cos\delta\right] \tag{8}$$

It may be seen that the contrast of the fringes can be controlled by the reference beam intensity during reconstruction. In general, the contrast is less than one, but by appropriate increase of the intensity of the reference beam, it is possible to achieve unit contrast fringes. In this particular example, if the bias transmittance t_0 is equal to $2\beta TR^2$, unit contrast fringe pattern results. Further, dark fringes are formed wherever $\delta = 2m\pi$, m being an integer.

B. Double-Exposure HI

Here both the exposures belonging to the initial and final states of the object are recorded sequentially on the same photographic plate. The total exposure recorded can be expressed as follows:

$$\beta T\{I_1(x, y) + I_2(x, y)\} \tag{9}$$

The transmittance of the hologram is expressed as follows:

$$\begin{aligned} t(x, y) = t_0' - \beta T[2O_o^2(x, y) &+ \{\mathbf{O}(x, y) + \mathbf{O}(x, y)e^{i\delta}\}\mathbf{R}^*(x, y) \\ &+ \{\mathbf{O}^*(x, y) + \mathbf{O}^*(x, y)e^{-i\delta}\mathbf{R}(x, y)] \end{aligned} \tag{10}$$

where $t_0' = t_0 - 2\beta TR^2$. When the double-exposure hologram is reconstructed, both the object waves are simultaneously released, producing on interference a unit contrast fringe pattern characteristic of the deformation. The intensity distribution in the fringe pattern is

$$I = 2(\beta TR^2)^2 O_o^2(1 + \cos\delta) = I_o(1 + \cos\delta) \tag{11}$$

The technique is also known as frozen-fringe or lapsed-time HI. It may be noted that this technique compares only two states of the object. Figure 4 shows a double-exposure interferogram of a pipe with a defect. The pipe was loaded by an application of hydraulic pressure between exposures. It is seen that the defect is a region of thinner wall thickness. Table 2 compares real-time and double-exposure HI.

XI. FRINGE FORMATION AND MEASUREMENT OF DISPLACEMENT VECTOR

We have mentioned that object deformation causes a phase change between the two waves, at least one of which is derived from the

Figure 4 Double-exposure interferogram of a pipe with a defect.

hologram. This phase change is responsible for the formation of the fringe pattern. In this section we address two issues:

How is the phase change related to the deformation vector?
Where are the fringes really localized?

Let us consider a point P on the object surface. Loading causes this point to move to a different location P'. The vector distance PP' is the deformation vector $\mathbf{d}$. We consider the geometry shown in Figure 5 to calculate the phase difference introduced by the deformation. The phase difference δ is given by

$$\delta = \mathbf{k}_1 \cdot \mathbf{r}_1 + \mathbf{k}_2 \cdot \mathbf{r}_2 - (\mathbf{k}_1 + \Delta\mathbf{k}_1) \cdot \mathbf{r}_1' - (\mathbf{k}_2 + \Delta\mathbf{k}_2) \cdot \mathbf{r}_2' \tag{12}$$

We can express $\mathbf{r}_1'$ and $\mathbf{r}_2'$ as follows:

$$\mathbf{r}_1 + \mathbf{d} = \mathbf{r}_1' \tag{13a}$$

$$\mathbf{r}_2' + \mathbf{d} = \mathbf{r}_2. \tag{13b}$$

Substituting for $\mathbf{r}_1'$ and $\mathbf{r}_2'$, we obtain

Table 2 Comparison of Single- and Double-Exposure HI

Single-exposure HI	Double-exposure HI
Monitors/compares different states of the object (due to loading) with the initial state over a range determined by usable fringe density and speckle noise.	Compares only final state with the initial state.
In situ processing is recommended; otherwise the hologram must be repositioned very accurately. Once an experimental setup has been disturbed, the hologram cannot be used for comparison.	Hologram can be reconstructed at leisure, even at a different location.
Fringe contrast is poor but can be improved.	Unit contrast fringe pattern is obtained as the diffraction efficiency is equally shared by the two beams.
Bright fringes are formed when $\delta = (2m+1)\pi$. Development produces a phase change of π.	Bright fringes are formed when $\delta = 2m\pi$.
Usually performed to obtain right parameters for double-exposure HI.	Usually used for quantitative evaluation.

$$\begin{aligned}\delta &= \mathbf{k}_1 \cdot \mathbf{r}_1 + \mathbf{k}_2 \cdot \mathbf{r}_2 - (\mathbf{k}_1 + \Delta\mathbf{k}_1) \cdot (\mathbf{r}_1 + \mathbf{d}) - (\mathbf{k}_2 + \Delta\mathbf{k}_2)(\mathbf{r}_2 - \mathbf{d}) \\ &= \mathbf{k}_1 \cdot \mathbf{r}_1 + \mathbf{k}_2 \cdot \mathbf{r}_2 - \mathbf{k}_1 \cdot \mathbf{r}_1 - \Delta\mathbf{k}_1 \cdot \mathbf{r}_1 - \mathbf{k}_1 \cdot \mathbf{d} - \Delta\mathbf{k}_1 \cdot \mathbf{d} - \mathbf{k}_2 \cdot \mathbf{r}_2 \\ &\quad + \mathbf{k}_2 \cdot \mathbf{d} + \Delta\mathbf{k}_2 \cdot \mathbf{d} - \Delta\mathbf{k}_2 \cdot \mathbf{r}_2 \\ &= (\mathbf{k}_2 - \mathbf{k}_1) \cdot \mathbf{d}\end{aligned} \tag{14}$$

This expression was obtained under experimental conditions in which the displacement vector was of the order of a few micrometers and the distances r_1 and r_2 could be from a few tens of centimeters to meters. When this is taken into account, the vectors

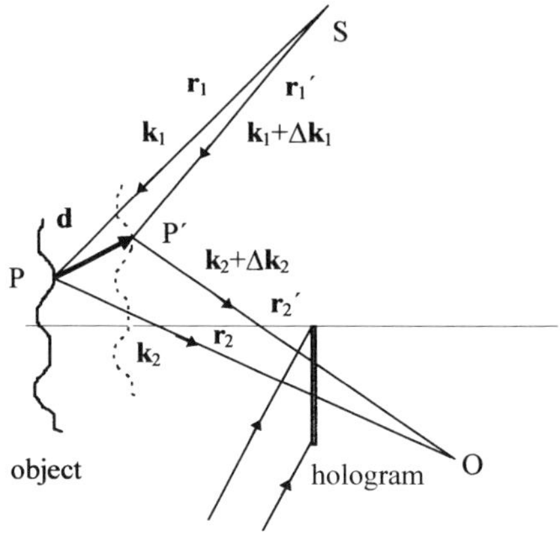

Figure 5 Calculation of the phase difference.

$\mathbf{r}_1$ and $\mathbf{r}_2$ are perpendicular to $\Delta\mathbf{k}_1$ and $\Delta\mathbf{k}_2$, respectively; hence their product terms are automatically zero, while the remaining two terms, $\Delta\mathbf{k}_1 \cdot \mathbf{d}$ and $\Delta\mathbf{k}_2 \cdot \mathbf{d}$, which represent second-order contributions, are also neglected.

This expression tells us only when and how the fringe pattern is formed; it does not tell us where the fringe pattern is actually formed. The situation is quite different from that encountered in classical interferometry. In HI, the fringes normally are localized not on the object surface but on a surface in space. This surface is known as the surface of localization. The knowledge of the location of the surface of localization helps in determining the deformation vector. In fact, the surface of localization depends on several factors, including the deformation vector. Several theories have been presented in the literature. Needless to say, a holographer should know about the fringe localization in detail before any quantitative evaluation is carried out.

XII. LOADING OF THE OBJECT

Since HI sees only the change in the state of the object, the object must be subjected to external agency to change its state. This can be performed by any one of the following methods:

1. Mechanical loading
2. Thermal loading
3. Pressure/vacuum loading
4. Vibratory or acoustical loading
5. Impact loading

The most commonly used method is mechanical loading, wherein the object is subjected to either compressive or tensile forces. To study the dynamic response, either vibratory or impact loading is imposed on the object. In holographic nondestructive testing (HNDT), it is very essential to use the appropriate kind of loading. For example, debonds are easily seen when pressure or vacuum stressing is applied. Table 3 gives some application areas for the five types of loading.

A. Time-Averaged HI

Time-averaged HI is utilized for the study of vibrating bodies like musical instruments. As the name suggests, the recording is carried out over a time period that is several times the period of vibration. Since an object vibrating sinusoidally spends most of the time at the locations of maximum displacement, a recording of such a vibrating object is equivalent to a double-exposure record. However, the intensity distribution in the reconstruction is modified considerably as a result of the time of excursion between these extreme positions. To study this phenomenon, we can write for the instantaneous intensity on the recording plane,

$$I(x, y; t) = O_o^2(x, y) + R^2 + \mathbf{O}^*(x, y; t)\mathbf{R}(x, y) + \mathbf{O}(x, y; t)\mathbf{R}^*(x, y) \tag{15}$$

where the object wave $\mathbf{O}(x, y; t)$ is expressed as follows:

$$\mathbf{O}(x, y; t) = O_o \exp[i\phi_o(x, y)] \exp[i\delta(x, y; t)] \tag{16}$$

The phase $\delta(x, y; t)$ represents the phase change introduced by the vibration. If the body is vibrating with an amplitude $A(x, y)$ at a frequency ω and is illuminated and observed along the directions

Table 3 Holographic Testing Applications

Problem detected	Items tested
Mechanical loading	
Cracks	In bolt holes in steel channels, concrete, rocket nozzle liners, welded metal plate, turbine blades, glass
Debonds	Honeycomb panels, propellant liners
Fatigue damage	Composite materials
Deformation	Radial deformation in carbon cylinders and tubes
Pneumatic loading	
Cracks	In welded joints, aluminum shells, bonded cylinders, ceramic heat exchanger tubes
Debonds/delamination	Composite panels and tubes, honeycomb structures, bonded cylinders, tires, rocket launch tubes, rocket nozzle liner
Weld defects	Welded plastic pipes, honeycomb panels
Weakness (thinness)	Aluminum cylinders, pressure vessles, tubes, composite tubes
Structural flaws	Composite domes
Solders joints	Medical implant devices
Bad brazes	Silicon pressure sensors
Thermal loading	
Cracks	In glass/ceramic tubes, train wheels, turbofan blades
Debonds	Aircraft wing assembly, honeycomb structures, laminated structures, rocket motor casing, rocket nozzle liners, rocket propellant liners, tires, bonded structures
Delaminations	Antique paintings
Various defects	Circuit boards and electronic modules
Vibrational loading	
Cracks	In train wheels, ceramic bars
Debonds	Turbine blades, rocket propellant liners, sheet metal sandwich structures, laminated structures, honeycomb structures, fiberglass-reinforced plastics, helicopter rotors, ceramics bonded to composite plates, brake discs, adhesive joints, metallic interfaces and elastomer
Strength	CRTs
Impulse loading	
Cracks	In steel plates, wing plank splices, turbine blades
Debonds	Foam insulation
Voids and thin areas	Aluminum plates

making angles θ_1 and θ_2 with the local normal, the phase difference δ can be written as follows:

$$\delta(x, y; t) = \frac{2\pi}{\lambda} A(x, y)(\cos\theta_1 + \cos\theta_2)\sin\omega t \tag{17}$$

The intensity distribution expressed by Eq. (15) is recorded over a period T much longer than the period of vibration; that is, intensity distribution is recorded over a large number of vibration cycles. The average intensity recorded over the period T is

$$I(x, y) = \frac{1}{T}\int_0^T I(x, y; t)\, dt \tag{18}$$

This record, on development, is called a time-averaged hologram, and the procedure is known as time-averaged holographic interferometry. The hologram is reconstructed with the reference wave. Since, on reconstruction, the various waves generated are separable, we consider only the amplitude of the desired wave that is,

$$a(x, y) = -\beta TR^2 \frac{1}{T}\int_0^T \mathbf{O}(x, y; t)\, dt = -\beta TR^2 \mathbf{O}(x, y) \frac{1}{T}\int_0^T \exp\left[\frac{2\pi i}{\lambda} A(x, y)(\cos\theta_1 + \cos\theta_2)\sin\omega t\right] dt \tag{19}$$

The time integral $(1/T)\int_0^T \exp[(2\pi i/\lambda)A(x, y)(\cos\theta_1 + \cos\theta_2) \sin\omega t]dt$ is called the characteristic function of the sinusoidal vibration, and is represented by M_T. Thus, the intensity distribution in the reconstructed image is given by

$$I_{rec}(x, y) = a(x, y)a^*(x, y) = \beta^2 T^2 R^4 O_o^2 |M_T|^2 \tag{20}$$

The characteristic function has been generalized for various motion types and is tabulated in Table 4 for motions of various kinds. The characteristic function for the sinusoidal motion as indicated by Eq. (17) can be analytically obtained. It is given by

Table 4 Characteristic Function $|M_T|^2$

HI type	Displacement	$\lvert M_T\rvert^2$
Real-time	Static (**L**)	$1 + c^2 - 2c\cos(\mathbf{k}\cdot\mathbf{L})$
	Harmonic of amplitude $\mathbf{A}(x, y)$	$1 + c^2 - 2cJ_0(\mathbf{k}\cdot\mathbf{A})$
	Harmonic (time-averaged)	$1 - J_0(\mathbf{k}\cdot\mathbf{A})$
Real-time with reference fringes	Harmonic	$1 + c^2 - 2c\cos(\mathbf{k}\cdot\mathbf{L})J_0(\mathbf{k}\cdot\mathbf{A})$
Real-time stroboscopic	Harmonic: pulses at $\omega t = \pi/2$ and $3\pi/2$	$1 + c^2 - 2c\cos(2\mathbf{k}\cdot\mathbf{A})$
Double-exposure	Static (**L**)	$\cos^2(\mathbf{k}\cdot\mathbf{L}/2)$
Double-exposure stroboscopic	Harmonic: pulses at $\omega t = \pi/2$ and $3\pi/2$	$\cos^2(\mathbf{k}\cdot\mathbf{A})$
Time-averaged	Harmonic of amplitude $\mathbf{A}(x, y)$	$J_0^2(\mathbf{k}\cdot\mathbf{A})$
	Constant velocity $\mathbf{L}_r = \mathbf{v}T$	$\mathrm{sinc}^2(\mathbf{k}\cdot\mathbf{L}_r/2)$
	Constant acceleration from rest	$\dfrac{C^2(\sqrt{\mathbf{k}\cdot\mathbf{A}}) + S^2(\sqrt{\mathbf{k}\cdot\mathbf{A}})}{(2/\pi)(\mathbf{k}\cdot\mathbf{A})}$
Time-averaged	Irrationally related modes	$J_0(\mathbf{k}\cdot\mathbf{A}_1)J_0(\mathbf{k}\cdot\mathbf{A}_2)$
Temporally frequency translated	Harmonic motion	$J_m^2(\mathbf{k}\cdot\mathbf{A})$
	Amplitude-modulated reference wave: $f_r(t) = e^{i(\omega t-\Delta)}$	$J_m^2(\mathbf{k}\cdot\mathbf{A})\cos^2\Delta$
	Phase-modulated reference wave: $f_r(t) = \exp[i(M_R\sin\omega t)]$	$J_m^2(\mathbf{k}\cdot\mathbf{A} - \mathbf{M}_R)$

Key: c measures the contrast, and C and S are Fresnel cosine and sine integrals.

$$
\begin{aligned}
M_T &= \frac{1}{T}\int_0^T \exp\left[\frac{2\pi i}{\lambda}A(x, y)(\cos\theta_1 + \cos\theta_2)\sin\omega t\right]dt \\
&= J_0\left[\frac{2\pi}{\lambda}A(x, y)(\cos\theta_1 + \cos\theta_2)\right]
\end{aligned}
\tag{21}
$$

where $J_0(x)$ is the Bessel function of zeroth order and first kind. The intensity distribution in the reconstructed image will be

$$I_{rec}(x, y) = I_0(x, y)J_0^2\left[\frac{2\pi}{\lambda}A(x, y)(\cos\theta_1 + \cos\theta_2)\right] \tag{22}$$

The variation of $J_0^2(x)$ with the argument x is shown in Figure 6. It may be seen that the reconstructed image is modulated by the $J_0^2(x)$ function. Thus, the image is covered by the fringes. Figure 7 shows the time-averaged interferogram of an edge-clamped diaphragm vibrating in the second harmonic mode. The zero intensity in the image occurs at the zeros of the $J_0^2(x)$ function. It may also be seen that the intensity is maximum at the stationary region $[A(x, y) = 0]$ and decreases to zero when

$$\frac{2\pi}{\lambda}A(x, y)(\cos\theta_1 + \cos\theta_2) = 2.4048 \tag{23}$$

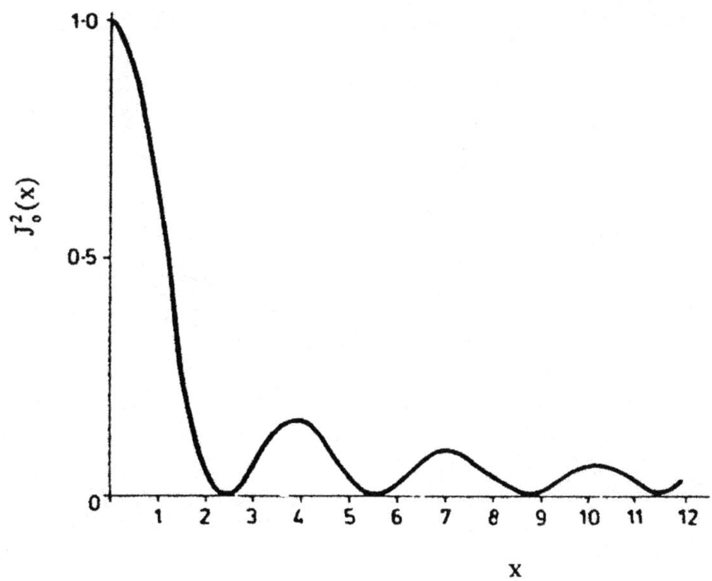

Figure 6 $J_0^2(x)$ distribution, which defines the intensity distribution in time-averaged HI fringes for a sinusoidally vibrating object.

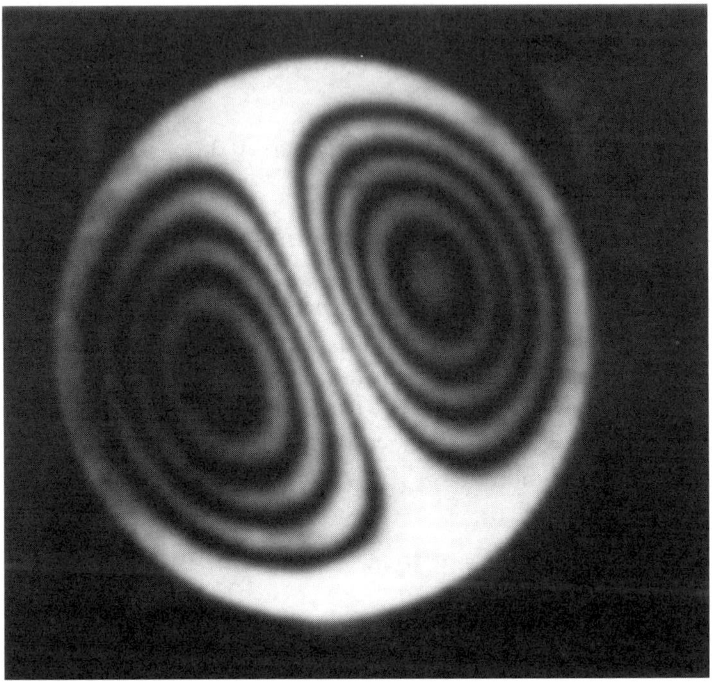

Figure 7 Time-averaged interferogram of an edge-clamped diaphragm vibrating in second harmonic mode. (From *Introduction to Fourier Optics* by Goodman, McGraw-Hill, 1968.)

This gives the amplitude of vibration at the first zero of the Bessel function. Assuming illumination and observation directions along the normal to the surface of the object, the amplitudes at the first and the successive zeros of the Bessel function are given by

$$A(x, y) = \frac{\lambda}{4\pi} 2.4048 \tag{24a}$$

$$= \frac{\lambda}{4\pi} 5.5200 \tag{24b}$$

$$= \frac{\lambda}{4\pi} 8.6537 \tag{24c}$$

$$= \frac{\lambda}{4\pi} 11.7915 \tag{24d}$$

Since the intensity falls off rapidly, high amplitudes of vibration are difficult to monitor by this technique. In brief, time-averaged HI gives the vibration map (phase of the vibration is lost) over a range of frequencies; that is, the amplitude of the vibration can be measured.

Let us consider a simple example of a cantilever shown in Figure 8. It is illuminated normally and observed in the same direction. We assume illumination by the red radiation of the He–Ne laser. The amplitude of vibration can be expressed as follows:

$$z(x, t) = z(x)\sin\omega t \tag{25}$$

The intensity distribution in the image is given by

$$I_{\mathrm{rec}}(x, y) = I_0(x, y)J_0^2\left[\frac{4\pi}{\lambda}A(x, y)\right] \tag{26}$$

The fringes run parallel to each other, and for higher amplitudes are almost equidistant. Dark fringes are located where the amplitudes of vibration are 0.12 μm, 0.28 μm, 0.44 μm, 0.59 μm, ...

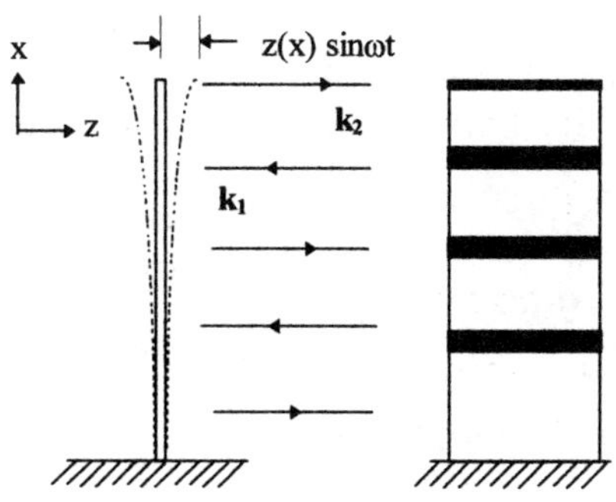

Figure 8 Time-averaged holographic interferometry of a vibration cantilever.

B. Real-Time, Time-Averaged HI

In real-time, time-averaged HI, a single-exposure record of a stationary object is first made. After development, the hologram is precisely relocated. The object is now set in vibration and viewed through the hologram. Interference between the wave reconstructed from the hologram and the directly transmitted wave from the vibrating object will yield an intensity distribution that is proportional to $[1 - J_0(x)]$, where $x = (4\pi/\lambda)A(x, y)$. Thus, the nodal points on the object appear dark. The contrast of this fringe pattern is very low. While one is observing the real-time pattern, the laser beam can be chopped at the frequency of vibration. The method is then equivalent to the real-time HI for static objects. The fringes correspond to the displacement between the initial position and the position at which the laser beam illuminates the object. By varying the time at which the light pulse illuminates the object during its vibration cycle, one can map the displacement at different phases of vibration.

XIII. MEASUREMENT OF VERY SMALL VIBRATION AMPLITUDES

It has been shown that the first minimum of intensity in time-averaged HI occurs when $A(x, y) = (\lambda/4\pi)2.4048$. This is valid only when the directions of illumination and observation are along the surface normal. For illumination from a He–Ne laser (λ = 632.8 nm) the amplitude of vibration $A(x, y)$ corresponds to 0.12 μm. Even smaller amplitudes of vibration can be monitored by observing the variation of intensity between the maximum and first zero of the Bessel function.

XIV. MEASUREMENT OF LARGE-VIBRATION AMPLITUDES

A. Frequency Modulation of the Reference Wave

The frequency of the reference wave is shifted by $n\omega$, where n is an integer and ω is the frequency of vibration, and a time-averaged hologram is recorded in the normal way. The recorded intensity distribution is given by

$$I(x, y) = \frac{1}{T}\int_0^T [O_o^2 + R_o^2 + \mathbf{OR}^* + \mathbf{O}^*\mathbf{R}]dt \tag{27}$$

where the object wave and the reference wave are written explicitly as follows:

$$\mathbf{O}(x, y; t) = O_o \exp\{i[\bar{\omega}t + \phi_o(x, y) + \delta(x, y; t)]\} \tag{28a}$$

and

$$\mathbf{R}(x, y; t) = R_o \exp\{i[(\bar{\omega} - n\omega)t + \phi_R]\} \tag{28b}$$

Further, assuming that the object is vibrating sinusoidally with an amplitude $A(x, y)$, we can write for the phase difference $\delta(x, y; t)$

$$\delta(x, y; t) = \frac{2\pi}{\lambda}(\cos\theta_1 + \cos\theta_2)A(x, y)\sin\omega t = \frac{2\pi}{\lambda}2A(x, y)\sin\omega t$$

when the directions of illumination and observation are along the surface normal. Since the various terms in the expression for the recorded intensity are separable, we pick up the term of interest as follows:

$$\begin{aligned} &= \frac{1}{T}\int_0^T \boldsymbol{OR}^*\, dt = O_o e^{i\phi_o} R_o e^{-i\phi_R} \frac{1}{T}\int_0^T \exp\left(\frac{4\pi}{\lambda} iA\sin\omega t\right) e^{-in\omega t} dt \\ &= O_o R_o e^{i(\phi_o - \phi_R)} \frac{1}{T}\sum_{n'=-\infty}^{n'=\infty} J_{n'}\left(\frac{4\pi}{\lambda}A\right)\int_0^T e^{in'\omega t} e^{-in\omega t}\, dt \end{aligned} \tag{29}$$

The integral will vanish for all values of n' except $n' = n$. Thus, the amplitude distribution, on reconstruction, becomes proportional to $J_n(4\pi/\lambda A)$. The intensity distribution is then given by

$$I(x, y) = I_0 J_n^2\left(\frac{4\pi}{\lambda} A\right) \tag{30}$$

Unlike the case of time-averaged HI, fringes of nearly equal intensity are formed; hence large vibration amplitudes can be measured.

B. Phase Modulation of the Reference Beam

The mirror in the reference arm is mounted on a PZT that is excited by a voltage signal at the frequency of vibration. The amplitude of the reference wave at the recording plane is

$$\mathbf{R}(x, y) = R_o \exp(i\phi_R) \exp(i\delta_R) \tag{31}$$

where $\delta_R = (2\pi/\lambda)2A_R \sin(\omega t - \Delta)$ with A_R and Δ representing the amplitude and phase of the reference mirror vibration. The object wave is expressed as follows

$$\mathbf{O}(x, y) = O_o \exp[i\phi_o(x, y)] \exp(i\delta) \tag{32}$$

where $\delta = (2\pi/\lambda)2A(x, y) \sin \omega t$ is the phase difference introduced by object vibration. It is to be noted that the modulation in the reference wave is phase-shifted by Δ, which may be experimentally varied. Now, a time-averaged hologram is recorded. Again, we consider only the term of interest in the expression for amplitude transmittance, and we express the amplitude of the reconstructed wave as follows:

$$\begin{aligned} \propto \frac{1}{T}\int_0^T \mathbf{O}\mathbf{R}^* dt &= O_o R_o \exp[i(\phi_o - \phi_R)] \\ &\quad \frac{1}{T}\int_0^T \exp\left\{\frac{4\pi}{\lambda} i[A \sin \omega t - A_R \sin(\omega t - \Delta)]\right\} dt \\ &= O_o R_o \exp[i(\phi_o - \phi_R)] \\ &\quad J_0\left[\frac{4\pi}{\lambda}(A^2 + A_R^2 - 2AA_R \cos \Delta)^{1/2}\right] \end{aligned} \tag{33}$$

The intensity distribution in the reconstructed image is given by

$$I(x,y) = I_0 J_0^2 \left\{ \frac{4\pi}{\lambda} \left(A^2 + A_R^2 - 2AA_R \cos\Delta \right)^{1/2} \right\} \tag{34}$$

As a special case, when the phase of modulation Δ of the reference wave is zero, the intensity distribution is given by

$$I(x,y) = I_0 J_0^2 \left\{ \frac{4\pi}{\lambda} (A - A_R) \right\} \tag{35}$$

Thus, control of the amplitude of the reference mirror vibration can be used to measure large vibration amplitudes. The zeroth-order fringe (the brightest maximum) will be formed at the regions in which the vibration amplitude of the reference mirror matches that of the object. In this way, it is possible to extend the measurable range considerably, in practice up to about 10 μm. By varying the phase of the reference mirror, it is also possible to trace out areas of the object vibrating in the same phase as that of the reference mirror, thereby mapping the phase distribution of the object vibration.

XV. STROBOSCOPIC ILLUMINATION/ STROBOSCOPIC HI

The object is illuminated by laser pulses whose duration is much shorter than the period of vibration. The recorded hologram is thus like a double-exposure hologram. Let the first record be made when the object is in a state of vibration 1 and the second record be made when the object is in state 2 as shown in Figure 9. The double-exposure hologram will display fringes corresponding to the displacement A between these two vibration states. By varying the pulse separation, one can record holograms for different displacements.

Alternately, pulse separation may be adjusted to correspond to one-quarter of the time period, and the double-exposure holograms may be recorded at various phases of the vibration cycle. The advantage of stroboscopic HI is that fringes of unit contrast,

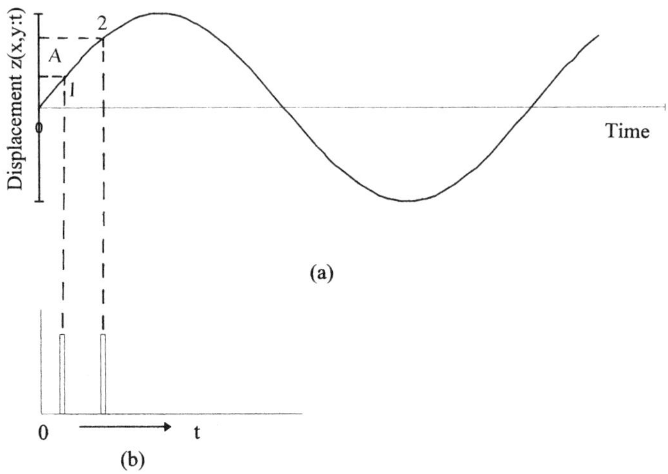

Figure 9 Stroboscopic HI of a vibrating object: (a) sinusoidally vibrating object and (b) pulse illumination.

similar to those in double-exposure HI, are formed. In addition, the use of pulse illumination does not require vibration isolation of the holographic setup.

XVI. SPECIAL TECHNIQUES IN HOLOGRAPHIC INTERFEROMETRY

A. Two-Reference-Beam HI

In the discussion of the phase shift method of evaluation of the phase in an interferogram, we mentioned that an additional phase is to be introduced that is independent of the phase to be measured. In real-time HI, this phase was provided by shifting the phase of the reference wave during reconstruction. Such a shift cannot be accomplished in double-exposure HI, however, because the reconstruction of both the recorded waves is done with a single reference beam.

To have independent access to the two reconstructed waves, and to introduce the desired phase difference, a holographic setup with two reference waves is required (Fig. 10a). The first exposure of a

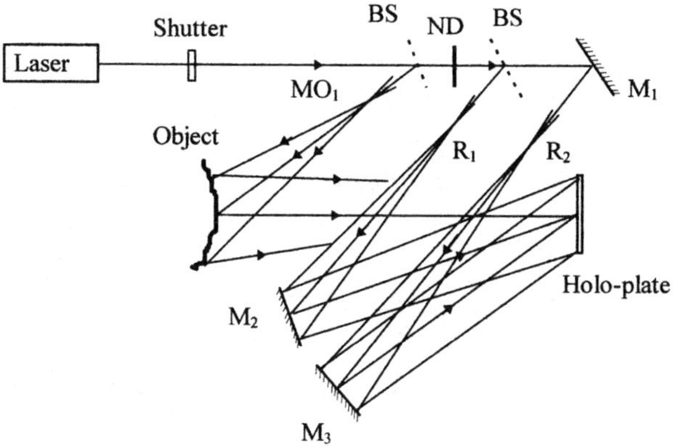

(a)

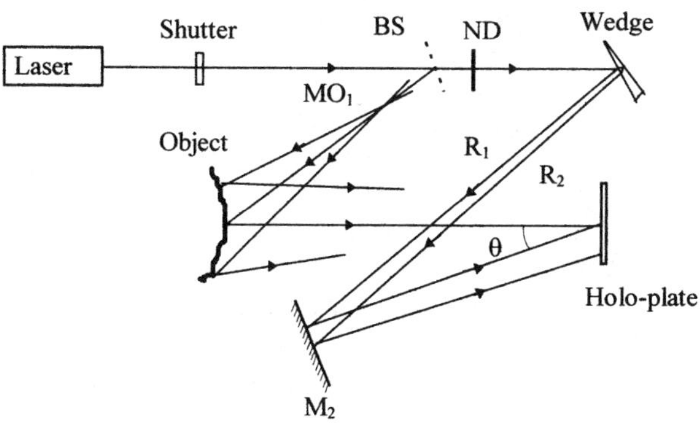

(b)

Figure 10 (a) Two-reference-waves HI. (b) Schematic for HI with two close reference waves.

double-exposure hologram is performed with the object in its initial state and with the reference wave R_1 (the reference wave R_2 is blocked). The object is loaded, and the second exposure is made with the reference wave R_2 (the reference wave R_1 is blocked). The reconstruction of the hologram is accomplished with R_1 and R_2 together to generate the interference pattern between the initial and final states of the object: the interference pattern may be varied by changing one of the reference waves while leaving the other unaffected. Mathematically, the procedure is described as follows.

In the first exposure, we record an intensity distribution $I_1(x, y)$, where

$$I_1(x, y) = O_o^2 + R_{10}^2 + \mathbf{O}\mathbf{R}_1^* + \mathbf{O}^*\mathbf{R}_1 \tag{36}$$

The intensity distribution recorded in the second exposure is given by

$$I_2(x, y) = O_0^2 + R_{20}^2 + \mathbf{O}'\mathbf{R}_2^* + \mathbf{O}'^*\mathbf{R}_2 \tag{37}$$

where $\mathbf{O}'(x, y) = \mathbf{O}(x, y)e^{i\delta(x,y)}$, being the object wave from the deformed state of the object. The total exposure is $T(I_1 + I_2)$, and the amplitude transmittance $t(x, y)$ of the doubly exposed hologram is

$$t(x, y) = t_0 - \beta T(I_1 + I_2) \tag{38}$$

This hologram is illuminated by both reference waves; therefore, the amplitude of the waves just behind the hologram is $(R_1 + R_2)\ t(x, y)$. This hologram generates a multiplicity of images that may or may not overlap. However, $\mathbf{R}_1\mathbf{R}_1^*\mathbf{O}$ and $\mathbf{R}_2\mathbf{R}_2^*\mathbf{O}'$ completely overlap and give rise to interference fringes. Other undesired images can be separated from this interference pattern if the angle between the two reference waves is made very large. Generally, two reference waves are taken on the same side of the object normal and have mutual angular separation much larger than the angular size of the object.

Though the desired images are separated by the large angular separation between the reference waves, the arrangement is highly sensitive to the misalignment of hologram during repositioning and also to any wavelength change between hologram recording and reconstruction. Assuming that the reference waves are plane waves propagating in directions given by propagation vectors $\mathbf{k}$

and $\mathbf{k}'$, the additional phase difference $\Delta\phi$ at any point $\mathbf{r}_H$ on the hologram due to misalignment and wavelength change is

$$\Delta\phi(\mathbf{r}_H) = [(\mathbf{k} - \mathbf{k}') \times \boldsymbol{\omega}] \cdot \mathbf{r}_H + \frac{\Delta\lambda}{\lambda}(\mathbf{k} - \mathbf{k}') \cdot \mathbf{r}_H \tag{39}$$

where $\boldsymbol{\omega}$ $(= \Delta\xi, \Delta\eta, \Delta\chi)$ is the rotation vector for small rotations $\Delta\xi$, $\Delta\eta$, $\Delta\chi$ of the hologram about the x, y, and z axes, and $\Delta\lambda$ is the wavelength change between recording and reconstruction. It is seen that the additional phase change $\Delta\phi$ will be small if $(\mathbf{k} - \mathbf{k}')$ is small (i.e., both reference waves propagate nearly in the same direction). This relatively small phase change comes at the price of a reduction of the fringe contrast, since all the reconstructed waves now overlap. A schematic of an arrangement to produce two close reference waves is shown in Figure 10b. This setup reduces the influence of misalignment errors and wavelength change error. In fact, if the recording and the reconstruction are done at the same wavelength, there is no error due to wavelength change. However, if the recording is done with a pulse laser and the reconstruction by a continuous wave laser, the error due to the wavelength change cannot be removed, although it can be reduced by means of the experimental arrangement just described.

B. Sandwich HI

HI records phase difference no matter what its source. However, when one is applying HI for studying strains in the object, the main interest is to measure local deformations on the object surface. Depending on the loading conditions, the object very often undergoes rigid body movements (translations and tilts), and these are commonly much greater in magnitude than the local deformations. These rigid body deformations might be too large to be measured by HI, or they may completely mask the local deformations of interest. Therefore, there is considerable interest in exploring techniques that can compensate for the influence of rigid body movements. Fringe control techniques have been developed, but they are applicable only with real-time HI. Sandwich HI offers an attractive alternative for compensating rigid body movements.

As the name suggests, the recording is done on two holographic plates H_1 and H_2, which form a sandwich (Fig. 11a). We assume a plane reference wave. The initial state of the object is recorded on H_1 (H_2 is replaced by a dummy), and the final state on H_2. A ray from an object point P hits the plate H_1 at B and plate H_2 at A. Because of this geometry, an additional path difference $(AB - BD)$ is recorded. From Figure 11a, we write $AB = s/\cos\alpha$ and $BD = AB\cos(\alpha + \beta)$ where s is the separation between the plates of the sandwich and α and β are the angles shown in Figure 11a. Therefore, the path difference $(AB - BD)$ is equal to $(s/\cos\alpha)[1 - \cos(\alpha + \beta)]$.

That is, if only the initial state of the object is recorded on both the plates, the reconstruction will give a fringe-free image of the object. But the path difference $(AB - BD)$ can be varied by tilting the sandwich, resulting in an appearance of a fringe pattern. We will now see how much change in the path difference is introduced by the tilt of the sandwich. Let us consider that the sandwich is tilted by a small angle ϕ so that the reference beam now makes an angle of β' with the normal to the sandwich, as shown in Figure 11b. The corresponding points on the tilted sandwich are designated as A_1 and B_1. The reconstructed rays from A_1 and B_1 are shown by the dotted line. The path difference between these rays is $[r\cos\phi - r\cos(\alpha + \beta')]$ with $\beta' = \beta - \phi$ and $r = s/\cos\alpha$. Therefore, the net path difference is

$$\begin{aligned} &r[\cos\phi - \cos(\alpha + \beta') - 1 + \cos(\alpha + \beta)] \\ &\cong r\phi\sin(\alpha + \beta) = s[\phi\tan(\alpha + \beta)] \end{aligned} \tag{40}$$

It is seen that the path difference change due to the tilt of the sandwich is linearly proportional to ϕ.

Let us now consider that as a result of loading, the object is tilted by a small angle $\Delta\xi$ as shown in Figure 11c. This introduces an additional path equal to $2x\Delta\xi$. For this tilt to be compensated, we have

$$2x\Delta\xi = s\phi\tan(\alpha + \beta) \tag{41}$$

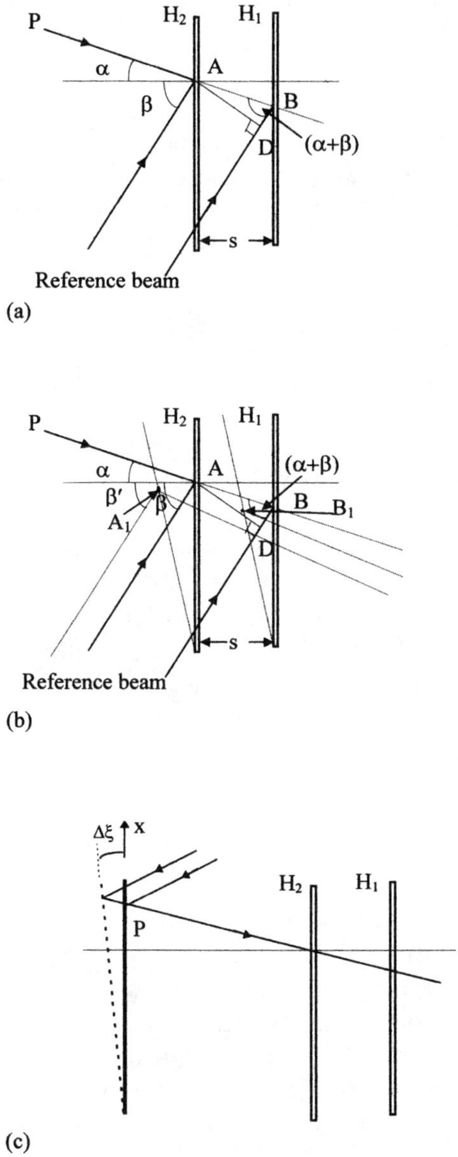

Figure 11 Sandwich hologram interferometry: (a) recording of a sandwich hologram, (b) influence of tilt of the sandwich, and (c) sandwich HI with a tilted object.

When the angles α and β are small and are also taken equal to each other (i.e., $\alpha = \beta$), Eq. (41) becomes

$$2x\Delta\xi = s\phi 2\alpha \tag{42}$$

But the angle α can be expressed as $\alpha = x/z$, where z is the distance between the sandwich hologram and the object plane. Therefore

$$\phi = \frac{z}{s}\Delta\xi \tag{43}$$

This magnitude of tilt ϕ of the sandwich hologram will fully compensate a tilt of magnitude $\Delta\xi$ of the object.

To fully utilize the strength of sandwich HI, a reference surface is placed by the side of the actual object—the reference surface should be free from fringes on reconstruction. When the fringe pattern on the object is compensated, however, a linear fringe pattern appears on the reference surface that is used to determine the tilt given to the sandwich hologram. Alternately, the reference surface is so placed that it is unaffected by loading but is influenced by the bodily tilts and translations. Then, on reconstruction, the reference surface will also carry a fringe pattern that should be fully compensated if one is to obtain the correct deformation map of the object.

C. Reflection HI

Another method to eliminate the contribution of rigid body movements and tilts is to clamp the holographic plate to the object under study, as shown in Figure 12a. The photographic plate, with its emulsion side facing the object, is illuminated by a plane wave at normal incidence. In the first pass through the plate, this wave acts as a reference wave, while the light scattered from the object forms the object wave. The reflection hologram is recorded within the emulsion as interference planes between the incident wave and the one scattered by the object. The recording is done on a thick emulsion, say of about 10 μm thickness: Agfa-Gevaert 8E75 plates are suitable for recording with He–Ne lasers.

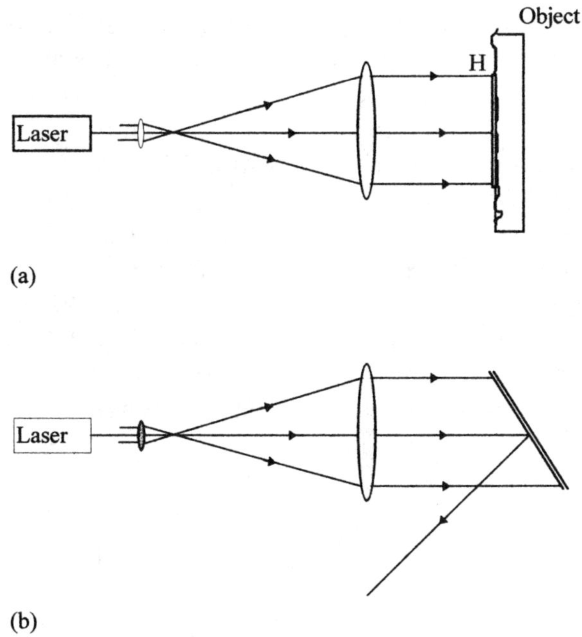

Figure 12 (a) Recording of a hologram in reflection HI and (b) its reconstruction.

Interferometric comparison requires two exposures, with the object loaded in between. The hologram is reconstructed with the laser light, as shown in Figure 12b. Since there is emulsion shrinkage, the reconstruction is possible only at a suitable lower wavelength. Otherwise, the emulsion is swelled to its original thickness by soaking it in an aqueous solution of triethanolamine $[(CH_2OHCH_2)_3N]$ followed by slow and careful drying in air to observe the fringe pattern at the recording wavelength.

If plane wave illumination is used during recording, the resulting hologram is insensitive to rigid body translations parallel and perpendicular to the direction of illumination. Also, in-plane rotation, for this case, has no influence. This is because either no path difference or a constant path difference is introduced. However, additional fringes may be introduced if the rigid body rotation is around an axis perpendicular to the direction of illumination and the object and the holographic plate are separated, as shown in Figure 13. We take a point P on the object and consider recording, say, at point H of the hologram. As a result of the rotation, the hologram–object combination takes a new position: P moves to P' as the object rotates, and H moves to H' as the plate rotates along with the object. In the first exposure, the optical path recorded is $IP + PH$, while in the second recording it is $MP' + P'H'$. The optical path difference Δ is $IP + PH - MP' - P'H'$. But $PH = P'H'$. Therefore the path difference Δ is $IP - MP' = PH\cos\beta - P'H'\cos(\beta + \phi)$, where β is the direction of observation and ϕ is the angle of rotation. This may be rewritten as follows:

$$\begin{aligned}\Delta &= PH[\cos\beta - \cos(\beta+\phi)] = PH\cos\beta\left[1 - \frac{\cos(\beta+\phi)}{\cos\beta}\right] \\ &= z_{\rm p}\left[1 - \frac{\cos\beta - \phi\ \sin\beta}{\cos\beta}\right] \quad \text{since } \phi \ll 1 \\ &= z_{\rm p}\phi\tan\beta\end{aligned} \tag{44}$$

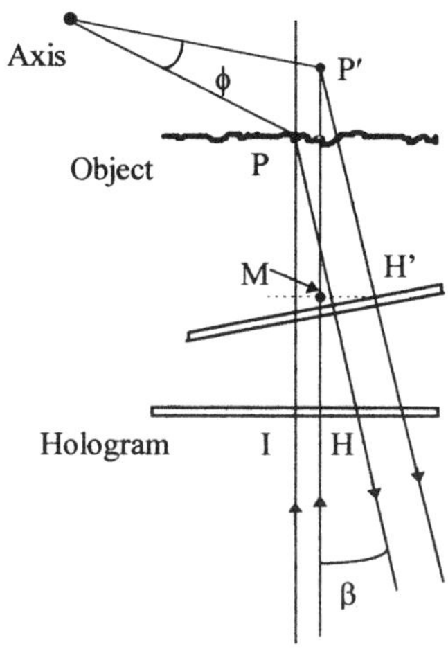

Figure 13 Calculation of path change when the hologram–object combination is rotated.

where z_p is the distance between the object and the holographic plate. It is seen that this path difference yields additional fringes. However, no additional fringes are introduced; that is, $\Delta = 0$ for $\phi \neq 0$, when z_p is zero, and/or when β is zero. Thus contacting the holographic plate with the object eliminates the influence of rotation about an axis perpendicular to the illumination beam.

XVII. EXTENDING THE SENSITIVITY OF HI: HETERODYNE HI

It has been shown that the fringe separation usually corresponds to $\lambda/2$ in path difference. With phase-shifting techniques, deformations as small as $\lambda/30$ can be measured. If even smaller deformations are to be monitored, other means are to be explored. Heterodyne HI is one such technique, being capable of giving resolution of $\lambda/1000$. This resolution is, however, achieved at the expense of more complicated experimental setup and slower data acquisition.

Figure 14 shows an experimental arrangement—a double-reference-beam geometry. One of the reference beams passes through two acoustooptical modulators. The first exposure is made with the reference beam R_1 (R_2 being blocked), and the second exposure,

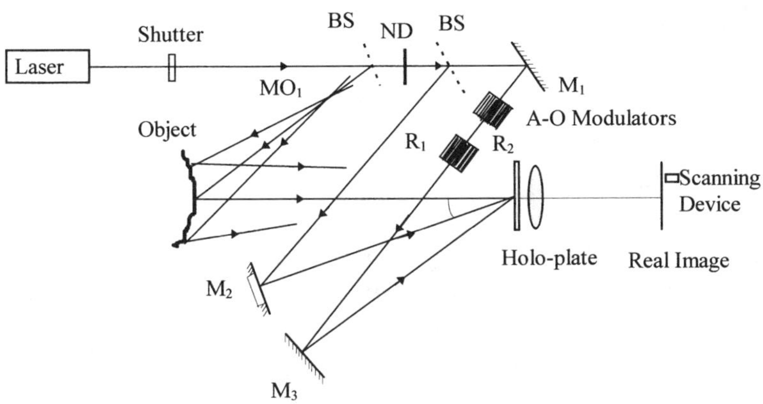

Figure 5.14 A schemmatic for the heterodyne hologram interferometry.

after the object has been loaded, is made with the reference wave R_2 (R_1 being blocked). Both exposures are made at the same wavelength with the AO modulators turned off. The reconstruction is done with both the reference waves; the AO modulators are turned on so that the wavelengths of the two reference waves are different. For example, one AO modulator modulates the beam at 40 MHz, and the frequency of the light wave diffracted in the first order is shifted by 40 MHz. This then passes through the second AO modulator, which modulates it at 40.08 MHz. We now consider the diffraction in -1 diffraction order to cancel the deviation of the beam, and obtain a light wave that is frequency-shifted by 80 kHz (= 40.08 – 40.0 MHz). Therefore, the two reference waves are now frequency-shifted by 80 kHz.

As pointed out earlier, a multiplicity of images are produced in two-reference-wave HI. We, however, consider only terms of interest in the amplitude transmittance of the hologram (i.e. $\mathbf{O}\mathbf{R}_1^* + \mathbf{O}'\mathbf{R}_2^*$, where reference waves $\mathbf{R}_1$ and $\mathbf{R}_2$ have the same frequency). For the reference and object waves, we explicitly write:

$$\mathbf{R}_1 = R_{01} \exp[i(\omega_1 t + \phi_{R_1})] \tag{45a}$$

$$\mathbf{R}_2 = R_{02} \exp[i(\omega_1 t + \phi_{R_2})] \tag{45b}$$

$$\mathbf{O} = O_o(x, y) \exp[i(\omega_1 t + \phi_o)] \tag{45c}$$

and

$$\mathbf{O}' = \mathbf{O} \exp[i\delta(x, y, z)] \tag{45d}$$

During reconstruction $\mathbf{R}_1 = R_{01} \exp[i(\omega_1 t + \phi_{R_1})]$ and $\mathbf{R}_2 = \mathbf{R}_2'$, where $\mathbf{R}_2' = R_{02} \exp[i(\omega_2 t + \phi_{R_2})]$, with $\omega_2 = \omega_1 \pm \Delta\omega$, $\Delta\omega$ is the frequency shift introduced by the AO modulators. Again writing for the terms of interest, the amplitude in the reconstructed image is proportional to

$$(\mathbf{R}_1 + \mathbf{R}_2')(\mathbf{O}\mathbf{R}_1^* + \mathbf{O}'\mathbf{R}_2^*)$$

Again with terms of interest under examination, we obtain the intensity distribution in the image as follows:

$$|\mathbf{OR}_1\mathbf{R}_1^* + \mathbf{O}'\mathbf{R}_2'\mathbf{R}_2^*|^2 = a(x, y) + 2b(x, y)\cos[\delta(x, y) + \Delta\omega t] \quad (46)$$

where the various appropriate terms have been absorbed in $a(x, y)$ and $b(x, y)$. If the intensity at a point (x, y) is observed, it is found to vary sinusoidally with time. The interference phase that is to be measured introduces a phase shift in this signal. Unfortunately, the phase $[\delta(x, y) + \Delta\omega t]$ carries no information. But from the phase difference of the oscillating intensities at two different points, one obtains

$$\begin{aligned}\Delta\delta(x_1, y_1 : x_2, y_2) &= [\delta(x_1, y_1) + \Delta\omega t] - [\delta(x_2, y_2) + \Delta\omega t] \\ &= \delta(x_1, y_1) - \delta(x_2, y_2)\end{aligned}$$

The phase difference $\Delta\delta$, which is the difference between the interference phases at two points, can be measured with a very high degree of accuracy by means of an electronic phasemeter.

There are two ways to perform the temporal heterodyne evaluation. In one approach, the phase difference is measured by keeping one detector fixed to a reference point and scanning the image with another detector. In this way, phase difference modulo 2π with respect to the phase subsisting at the reference point is measured. The interference phase difference can be summed up from point to point to yield the phase distribution along a line or a plane. In the second method, we scan the real image using a pair, a triplet, a quadruplet, or a quintuplet of photodetectors with known separation, and then measure the phase differences. In practice, the detector consists of a photodiode to which a fiber is pigtailed. The other end of the fiber scans the real image.

Heterodyne HI cannot be implemented as real-time HI because the experimental setup and also the surrounding environment require an extremely high degree of stability.

XVIII. HOLOGRAPHIC CONTOURING/SHAPE MEASUREMENT

For complete stress analysis, it is necessary to know the shape of the object under study. The following are three techniques to obtain the shape of an object:

change of wavelength between exposures: dual-wavelength method
change of refractive index of the medium surrounding the object between exposures: dual-refractive-index method
change of direction of illumination beam between exposures: dual-illumination method

A. Dual-Wavelength Method

We will describe dual-wavelength method in detail. The technique requires that a holographic recording of the object be made on the same plate with two slightly different wavelengths λ_1 and λ_2. The processed plate, the hologram, is then illuminated with the reference wave at wavelength λ_1 so that the original object wave is reconstructed at λ_1 together with the distorted reconstruction of the object recorded at wavelength λ_2.

The waves from original reconstruction and distorted reconstruction interfere to yield a fringe pattern that is related to the shape of the object. To understand the working of this principle, we again go through the mathematics of recording and reconstruction. In the first exposure, we record an intensity distribution that is proportional to

$$O_{\mathrm{o}}^2 + R_{01}^2 + \mathbf{O}_1\mathbf{R}_1^* + \mathbf{O}_1^*\mathbf{R}_1$$

The intensity recorded in the second exposure is proportional to

$$O_{\mathrm{o}}^2 + R_{02}^2 + \mathbf{O}_2\mathbf{R}_2^* + \mathbf{O}_2^*\mathbf{R}_2$$

where the object and reference waves are defined as follows:

$$\mathbf{O}_1 = O_0 \exp(i\mathbf{k}_1 \cdot \mathbf{r}) = O_0 \exp\left(\frac{2\pi}{\lambda_1} iz_0\right) \exp\left[\frac{2\pi}{2\lambda_1 z_0} i(x - x_0)^2\right] \tag{47a}$$

$$\mathbf{O}_2 = O_0 \exp(i\mathbf{k}_2 \cdot \mathbf{r}) = O_0 \exp(ik_2 z_0) \exp\left[\frac{k_2}{2z_0} i(x - x_0)^2\right] \tag{47b}$$

$$\mathbf{R}_1 = R_0 \exp(ik_1 x \sin\theta) \tag{47c}$$

and

$$\mathbf{R}_2 = R_0 \exp(ik_2 x \sin\theta) \tag{47d}$$

For reconstruction, the hologram is illuminated with a reference wave at wavelength λ_1. Therefore, considering the terms of interest, the amplitude distribution in the virtual image is

$$\alpha \quad \mathbf{R}_1(\mathbf{O}_1\mathbf{R}_1^* + \mathbf{O}_2\mathbf{R}_2^*) = \mathbf{O}_1 R_o^2 + \mathbf{O}_2 R_o^2 \exp[ix\sin\theta(k_1 - k_2)] \tag{48}$$

The term $\exp[ix\sin\theta(k_1 - k_2)]$ represents a different direction of propagation of the object wave $\mathbf{O}_2$. This phase term will be zero, provided the angle of the reference beam R_2, before the second exposure, is adjusted such that $k_1 \sin\theta = k_2 \sin\phi$, where ϕ is the angle the reference beam R_2 makes with the optical axis. Then, the amplitude distribution is rewritten proportional to

$$R_o^2 O_o \left\{1 + \exp[i(k_2 - k_1)z_0] \exp\frac{i(k_2 - k_1)}{2z_0}(x - x_0)^2\right\}$$

The phase term $[(k_2 - k_1)/2z_0](x - x_0)^2$ is very small because $x \ll z_0$ and $x_0 \ll z_0$, and hence the amplitude distribution in the reconstructed virtual image is

$$R_o^2 O_o\{1 + \exp[i(k_2 - k_1)z_0]\}$$

The intensity distribution in the image can therefore be written as follows:

$$I = I_0[1 + \cos(k_2 - k_1)z_0] = I_0\left[1 + \cos\frac{2\pi|\lambda_1 - \lambda_2|}{\lambda_1\lambda_2}z_0\right] \tag{49}$$

Bright fringes are formed wherever

$$\frac{2\pi|\lambda_1 - \lambda_2|}{\lambda_1\lambda_2}z_0(x, y) = 2m\pi$$

or

$$z_0(x, y) = \frac{m\lambda_1\lambda_2}{|\lambda_1 - \lambda_2|} \tag{50}$$

The contour interval is $\Delta z_0(x, y) = \lambda_1\lambda_2/|\lambda_1 - \lambda_2|$. Essentially, the method produces interference planes perpendicular to the z axis and with a separation of $\lambda_1\lambda_2/|\lambda_1 - \lambda_2|$. The sensitivity of the contour fringes can be varied from 1 μm to several millimeters using two suitable wavelengths. Figure 15 is an interferogram of an object taken with two wavelengths. The object is a spherical surface; hence the contour lines are circular.

B. Dual-Refractive-Index Method

Similar results are obtained when double-exposure HI is performed with a change in the refractive index of the medium surrounding the object. The object is placed in an enclosure with a transparent window and the first exposure made. The refractive index of the medium surrounding the object is changed, and then the second exposure is made on the same plate. The refractive index can be easily changed when, say, the water medium around the object is replaced by a mixture of water and alcohol. The hologram, when viewed, displays the object covered with interference planes.

If the vacuum wavelength of the recording light is λ_0, then the wavelength in a medium of refractive index n_1 is $\lambda_1 = \lambda_0/n_1$, and in the medium of refractive index n_2 is $\lambda_2 = \lambda_0/n_2$. Essentially, this is a two-wavelength HI, and the contour interval should be given by

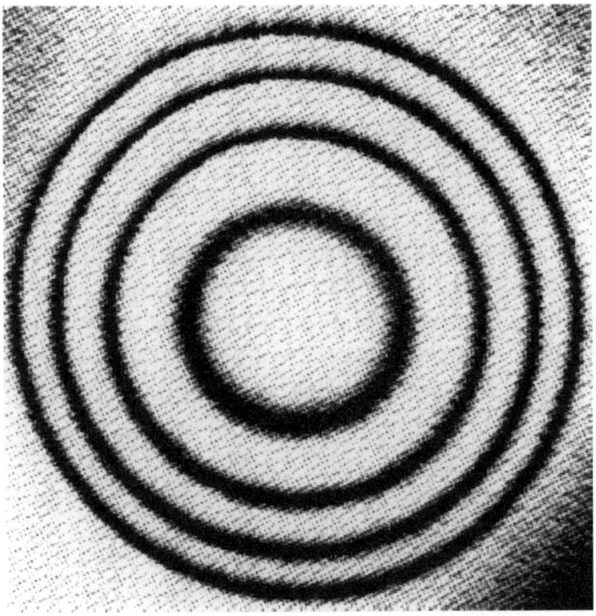

Figure 15 Contour fringes over a spherical surface obtained with dual-wavelength HI.

$$\Delta z_0(x, y) = \frac{\lambda_1 \lambda_2}{|\lambda_1 - \lambda_2|} = \frac{\lambda_0^2 / n_1 n_2}{|\lambda_0/n_1 - \lambda_0/n_2|} = \frac{\lambda_0}{|n_2 - n_1|} \tag{51}$$

There is no need to correct for the angle of the reference wave prior to the second exposure, as was done in two-wavelength HI. The contour interval can be varied by a suitable choice of the media surrounding the object.

C. Dual-Illumination Method

Dual illumination is frequently used for contouring because the method is simple. A collimated beam illuminates the object at an angle θ with the optical axis, and a real-time hologram is recorded in the usual way. On reconstruction, a dark field is observed. Now if the illumination beam is tilted by a small angle $\Delta\theta$, a system of equispaced interference planes intersects the object. These planes run parallel to the bisector of the angle $\Delta\theta$. The contour interval $\Delta z_0(x, y)$ is given by

$$\Delta z_0(x, y) = \frac{\lambda}{\sin\theta \, \sin\Delta\theta} \approx \frac{\lambda}{\sin\theta} \frac{1}{\Delta\theta} \tag{52}$$

Usually, the interference planes do not intersect the object perpendicular to the line of sight. Large objects are illuminated by a spherical wave from a point source that is translated laterally to produce the interference surfaces. Double-exposure HI can also be performed for contouring by shifting the point source between exposures.

XIX. HOLOGRAPHIC PHOTOELASTICITY

Holographic interferometry is widely applied to the study of photoelastic models. The technique is described in detail in Chapter 7 on photoelasticity.

XX. DIGITAL HOLOGRAPHY

A. Recording

Instead of using photoemulsions or other recording media, a CCD (charge-coupled device) is used as a recording medium and the information is stored electronically. The reconstruction is performed numerically on the electronically stored data. With CCDs, the recording of the hologram is done at video frequency, and no chemical or physical process for development is necessary. CCD cameras, however, have resolutions of ~100 l/mm, which is at least one order of magnitude smaller than that of photographic emulsions commonly used in holography. For this reason, the maximum admissible angle between the object and the reference waves is $\sim 1^\circ$.

A CCD contains $N \times N$ pixels of sizes Δx and Δy along the x and y directions. Assuming that the adjacent pixels have no spacing in between and no overlap, Δx and Δy are also pixel center distances. In digital holography, a Fresnel hologram is generated on the CCD target by the superposition of an object wave and a reference wave. The hologram is digitized, quantized, and stored in the memory of the image processing system. Figure 16 shows schematics for recording a digital hologram. A plane reference wave is assumed for simplicity. The reconstruction is done numerically. However, according to sampling theorem, only the spatial frequencies less than μ_{max} along the x direction and ν_{max} along the y direction can be reliably reconstructed where $\mu_{max} = 1/2\Delta x$ and $\nu_{max} = 1/2\Delta y$. This sets the maximum admissible angle between the object wave and the reference wave. The angle θ in the x–z plane can be written as $\theta = \sin^{-1}(\lambda\mu_{max})$. As a numerical example, consider a CCD of 1024×1024 pixels, each $6.8\ \mu m \times 6.8\ \mu m$ large, and the illumination of the object by a He–Ne laser beam at $0.633\ \mu m$. The maximum admissible angle is 2.67°. Therefore, an object placed at a distance of 1 m from the CCD must be smaller than 4.3 cm along the x direction. Larger objects must either be placed further away from the CCD plane or have their apparent size reduced by a lens.

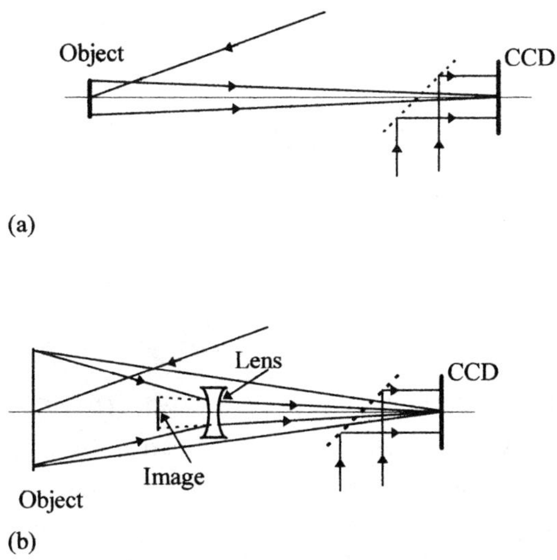

Figure 16 Schematic arrangements for recording a digital hologram: (a) small object and (b) large object.

B. Reconstruction

The hologram is reconstructed by illuminating it with a plane reference wave. The amplitude in the real image can be expressed as follows:

$$A(\xi, \eta) = \frac{iR}{\lambda z} \exp\left[-i\frac{\pi}{\lambda z}(\xi^2 + \eta^2)\right] \int_{-\infty}^{\infty} \int_{-\infty}^{\infty} t(x, y) \exp\left[-i\frac{\pi}{\lambda z}(x^2 + y^2)\right] \exp\left[\frac{2\pi i}{\lambda z}(x\xi + y\eta)\right] dx\, dy \qquad (53)$$

where R is the amplitude of the reference wave. This equation is valid under Fresnel approximation, which is

$$z^3 >> \frac{1}{8\lambda}\left[(x-\xi)^2 + (y-\eta)^2\right]^2$$

where z is the distance between the object and the hologram, hence between the hologram and the real image.

Since the data are available in digital form, Eq. (53) is written as follows:

$$A(m, n) = \frac{iR}{\lambda z} \exp\left[-i\frac{\pi}{\lambda z}(m^2 \Delta\xi^2 + n^2 \Delta\eta^2)\right] \sum_{k=0}^{N-1} \sum_{l=0}^{N-1} t(k, l) \exp\left[-i\frac{\pi}{\lambda z}(k^2 \Delta x^2 + l^2 \Delta y^2)\right] \exp\left[2\pi i\left(\frac{km}{N} + \frac{ln}{N}\right)\right] \qquad (54)$$

where $m = 0, 1, 2, \cdots (N-1)$; $n = 0, 1, 2 \cdots (N-1)$, and $t(k, l)$ is a matrix of $N \times N$ (pixels in CCD) data that describe the digitally sampled amplitude transmittance of the hologram; Δx, Δy and $\Delta\xi$, $\Delta\eta$ are the pixel sizes in the hologram plane and the plane of the real image, respectively. The amplitude distribution in the real image is obtained as the inverse Fourier transform of the product consisting of the hologram transmittance $t(k, l)$ multiplied by an exponential factor that contains a quadratic phase factor. The Fourier transform is performed by a fast Fourier transform algorithm. The amplitude $A(m, n)$ in the reconstructed image is a complex function; hence the intensity and the phase are computed at each pixel. The intensity $I(m, n)$ and the phase $\phi(m, n)$ are determined as follows:

$$I(m, n) = |A(m, n)|^2 = Re\{A(m, n)\}^2 + Im\{A(m, n)\}^2 \tag{55a}$$

and

$$\phi(m, n) = \tan^{-1} \frac{\mathrm{Im}\, A(m, n)}{\mathrm{Re}\, A(m, n)} \qquad \text{with value} \quad -\pi \text{ to } \pi \tag{55b}$$

As a consequence of surface roughness, the phase $\phi(m, n)$ varies randomly. In digital holography, only intensity variation on the object is of interest; hence several calculations like that of $iR/\lambda z$, phase $\phi(m, n)$ and the exponential term $\exp[-i(\pi/\lambda z(m^2 \Delta\xi^2 + n^2 \Delta\eta^2)]$ are not carried out.

XXI. DIGITAL HOLOGRAPHIC INTERFEROMETRY

Comparison of states of the object due to loading can be performed by digital holographic interferometry in a manner similar to that of double-exposure HI. Two holograms with the amplitude transmittances $t_1(x, y)$ and $t_2(x, y)$ belonging to the two states of the object are digitally recorded and stored. During reconstruction, we have two procedures to obtain the interferogram representing the change in the state of the object. In the first procedure, the transmittances of both the holograms are added pixel by pixel, and the Fresnel transform is taken. This will allow us to reconstruct the sum of the two amplitudes. When intensity distribution is calculated, it will exhibit cosine variation characteristics of the interference between the two waves. This interference pattern can be evaluated by any one of the well-known procedures. In the second method, the holograms are individually reconstructed and their phases $\phi_1(m, n)$ and $\phi_2(m, n)$ are calculated pixelwise. Though $\phi_1(m, n)$ and $\phi_2(m, n)$ are random functions, their difference

$\delta(m, n)[= \phi_2(m, n) - \phi_1(m, n)]$ will be deterministic and will give the phase change due to loading. The phase change $\delta(m, n)$ is calculated as follows:

$$\begin{aligned} \delta(m, n) &= \phi_2(m, n) - \phi_1(m, n) \qquad \phi_2(m, n) \geq \phi_1(m, n) \qquad (56a) \\ &= \phi_2(m, n) - \phi_1(m, n) + 2\pi \qquad \phi_2(m, n) < \phi_1(m, n) \qquad (56b) \end{aligned}$$

These two approaches are summarized in Table 5. Figures 17a and 17b show the numerically reconstructed phase of an undeformed and a deformed object, respectively, and the difference between these two is shown in Figure 17c as an interferogram. Since digital

Table 5 Procedures for Obtaining Digital Holographic Interferograms

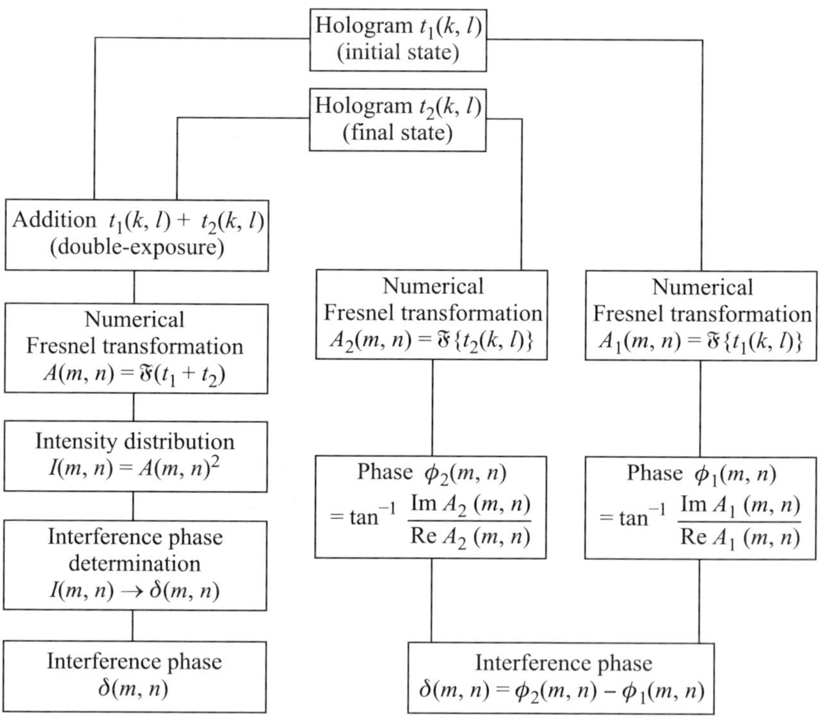

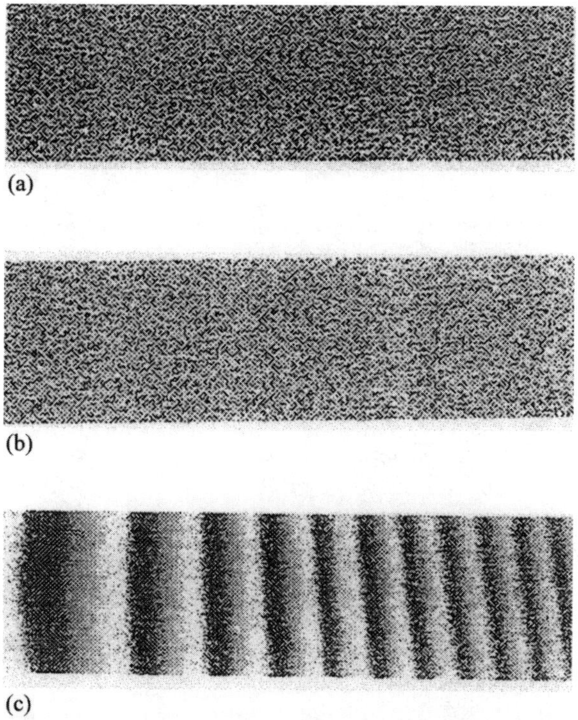

Figure 17 Numerically reconstructed phase of (a) undeformed object and (b) deformed object. (c) Phase difference between (a) and (b), producing an interferogram. (From Schnars, Kreis and Jüptner, *Opt. Eng.*, *35*(4), 977-982 (1996).)

HI can handle only small objects, it has potential applications in the testing and characterization of microsystems.

BIBLIOGRAPHY

1. R. K. Erf (Ed.), *Holographic Non-Destructive Testing*, Academic Press, New York (1974).
2. Ju. I. Ostrovskij, M. M. Butusov, and G. V. Ostrovskaja, *GolografiCeskaja Interferometrija* (in Russian), Nauka, Moscow (1977). Yu. I. Ostrovsky, M. M. Butusov, and G. V. Ostrovskaya, *Interferometry by Holography*, Springer Series in Optical Sciences, Vol. 20, Springer-Verlag, Berlin (1980).

3. C. M. Vest, *Holographic Interferometry*, Interscience, New York (1979).
4. W. Schumann and M. Dubas, *Holographic Interferometry*, Springer Series in Optical Sciences Vol.16, Springer-Verlag, Berlin (1979).
5. N. Abramson, *The Making and Evaluation of Holograms*, Academic Press, London (1981).
6. G. Wernicke and W. Osten, *Holografische Interferometrie* (in German), Physik-Verlag, Weinheim (1982).
7. R. J. Jones and C. Wykes, *Holographic and Speckle Interferometry*, Cambridge University Press, Cambridge (1983).
8. W. Schumann, J. P. Zürcher, and D. Cuche, *Holography and Deformation Analysis*, Springer Series in Optical Sciences Vol. 46, Springer-Verlag, Berlin (1985).
9. Yu. I. Ostrovsky, V. P. Schepinov, and V. V. Yakovlev, *Holographic Interferometry in Experimental Mechanics*, Springer Series in Optical Sciences Vol. 60, Springer-Verlag, Berlin (1991).
10. P. K. Rastogi (Ed.), *Holographic Interferometry, Principles and Methods*, Springer Series in Optical Sciences Vol. 68, Springer-Verlag, Berlin (1994).
11. Thomas Kreis, *Holographic Interferometry, Principles and Methods*, Akademie Verlag Series in Optical Metrology Vol. 1, Akademie Verlag, Berlin (1996).
12. Rajpal S. Sirohi and Klaus D. Hinsch, Selected Papers on *Holographic Interferometry*, Society of Photo Optical Instrumentation Engineers, Milestone Series MS144, SPIE Publishing, Bellingham, WA (1998).
13. Valery P. Shchepinova and Vladimir S. Pisarev, *Strain and Stress Analysis by Holographic and Speckle Interferometry*, John Wiley, Cheshire (1996).
14. Pramod K. Rastogi (Ed.), *Optical Measurement Techniques and Applications*, Artech House, Boston (1997).

Chapter 6

Speckle Metrology

I. THE SPECKLE PHENOMENON

Illumination of a diffuse object by coherent light produces a grainy structure in space. This grainy light distribution, known as speckle pattern, results from self-interference of numerous waves from scattering centers on the surface of the diffuse object as shown in Figure 1: the amplitudes and phases of these scattered waves are random variables. We assume that (1) the amplitude and the phase of each scattered wave are statistically independent variables, and also independent of the amplitudes and the phases of all other waves, and (2) the phases of these waves are uniformly distributed between $-\pi$ and π. Such a speckle pattern is fully developed. The resultant complex amplitude $\mathbf{u}(x, y) = u(x, y)e^{i\phi}$ is given by

$$u(x, y)e^{i\phi} = \frac{1}{\sqrt{N}}\sum_{k=1}^{N} u_k = \frac{1}{\sqrt{N}}\sum_{k=1}^{N} a_k e^{i\phi_k} \tag{1}$$

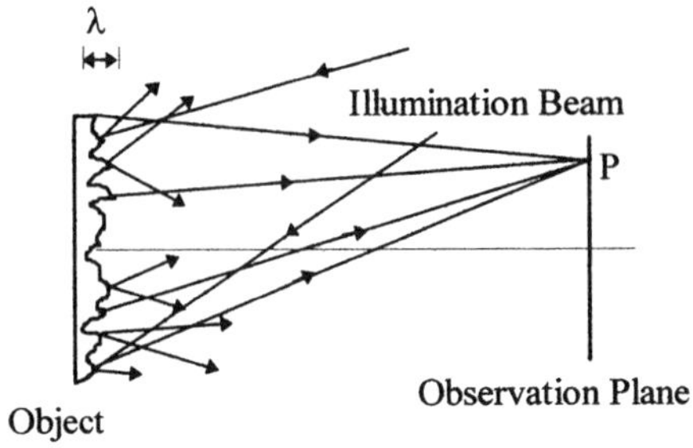

Figure 1 Formation of a speckle pattern.

where a_k and ϕ_k are the amplitude and the phase from the kth scatterer. For such a speckle pattern the complex amplitude of the resultant $\mathbf{u}(x, y)$ obeys the Gaussian statistics.

The probability density function $p(I)$ of speckle intensity is given by

$$p(I) = \frac{1}{\bar{I}} e^{-(I/\bar{I})} \tag{2}$$

where $\bar{I}$ is the average intensity. This gives the probability that the intensity at a point in the speckle pattern will have a value I. The probability density function in the speckle pattern follows a negative exponential law. The most probable intensity value is zero. A measure of the contrast in the speckle pattern is the ratio $c = \sigma/\bar{I}$, where σ is the standard deviation of the speckle intensity. The contrast in the fully-developed linearly polarized speckle pattern is unity.

II. AVERAGE SPECKLE SIZE

The grains or the speckles in the pattern are not well defined but have a structure. However, we associate with the speckle an average size. This is considered under the cases of objective and subjective speckle pattern.

A. Objective Speckle Pattern

It was pointed out earlier that the speckle pattern is formed in space when a diffuse object is illuminated by a coherent wave. This pattern, due to free space propagation, is termed an objective speckle pattern. The speckle size in an objective speckle pattern is given by

$$\sigma_{\text{ob}} = \frac{\lambda z}{D} \tag{3}$$

where D is the size of the illuminated area of the object and z is the distance between the object and the observation plane (Fig. 2a). The size is governed by the interference between the waves from the extreme scattering points. The relationship is the same as that expected from the Young's double-slit experiment, the slits being at the extreme positions in the illuminated area. The speckle size

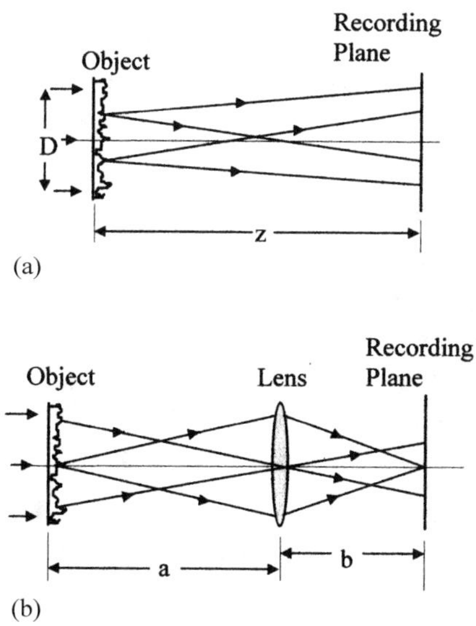

Figure 2 (a) Objective speckle pattern and (b) subjective speckle pattern.

increases linearly with the separation between the object and the observation plane.

B. Subjective Speckle Pattern

A speckle pattern formed at the image plane of a lens is called the subjective speckle pattern. It is due to interference of waves from several scattering centers in the resolution element (area) of the lens: in the image of this resolution area, the randomly dephased impulse response functions are added, resulting in a speckle. Therefore, the speckle size is governed by the well-known Airy formula. The diameter of the Airy disc is approximately given by

$$\sigma_s \cong \frac{\lambda b}{D'} \tag{4}$$

where D' is the diameter of the lens and b is the image distance (Fig. 2b). Here again, the speckle size is determined by the extreme size of the lens; the objective speckle pattern just in front of the lens is transmitted, hence is also obtained on the other side of the lens. The speckles at the extreme of this pattern determine the speckle size at the image plane. By introducing the f-number F# $(= f/D')$ of the lens, the average speckle size is expressed as follows:

$$\sigma_s = (1 + m)\lambda F\# \tag{5}$$

where we have introduced magnification $m\{= b/a = (b - f)/f\}$ of the lens. It is thus seen that the speckle size can be controlled by the magnification m and by the F# of the lens. The control of speckle size by F# is often used in speckle metrology to match the speckle size with the pixel size of the CCD array detectors.

III. SUPERPOSITION OF SPECKLE PATTERNS

The speckle patterns can be added either on amplitude or an intensity basis. An example of addition of speckle patterns on the amplitude basis is found in shear speckle interferometry, where the two speckle patterns are shifted and then superposed. In such a speckle pattern, the statistics of the speckle pattern remains unchanged.

However, when the speckle patterns are added on the intensity basis—for example, two speckle records are made on the same plate—the speckle statistics is completely modified and is governed by the correlation coefficient.

IV. SPECKLE PATTERN AND OBJECT SURFACE CHARACTERISTICS

The explanation of fully developed linearly polarized speckle pattern stressed that the phases are uniformly distributed between $-\pi$ and π and that large number of scatterers participate in the speckle formation. This results in unit contrast speckle pattern. However, if the surface is smoother than is necessary to satisfy this condition, the speckle contrast decreases. The speckle contrast depends on the surface roughness and coherence of light. In fact, the speckle contrast has been employed to measure the surface roughness over a large range using monochromatic and polychromatic illumination.

V. THE SPECKLE PATTERN AND SURFACE MOTION

A. Linear Motion in the Plane of the Surface

Let us assume a translucent object that is translated by a distance d in its own plane: the objective speckle pattern also translates by the same magnitude in the same direction (Fig. 3a). However, the structure of the speckle pattern begins to change (i.e., decorrelation sets in) when some of the scattering centers go out of the beam. For the subjective speckle pattern, speckle motion is in the direction opposite to that of the object and its magnitude is md, where m is the magnification (Fig. 3b).

B. Out-of-Plane Displacement

Let us consider that a speckle is formed at a position $P(r, 0)$ as shown in Figure 4a. This speckle is due to the superposition of all the waves from the object. When the object translates axially by a small amount ε, all these waves are phase-shifted nearly by the same amount. If this phase shift is a multiple of 2π, then a similar

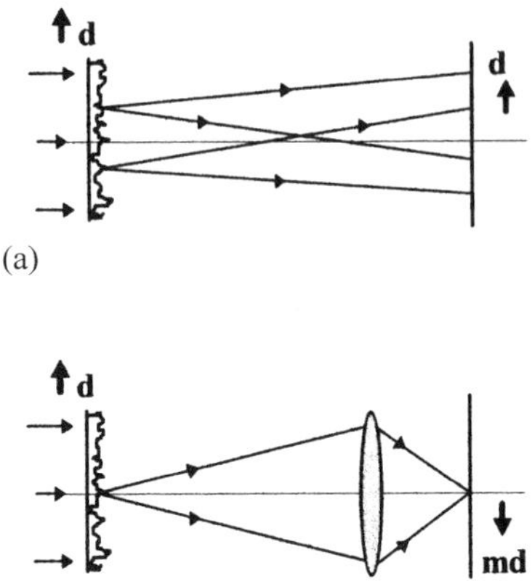

(a)

(b)

Figure 3 In-plane displacement: (a) objective speckle pattern and (b) subjective speckle pattern.

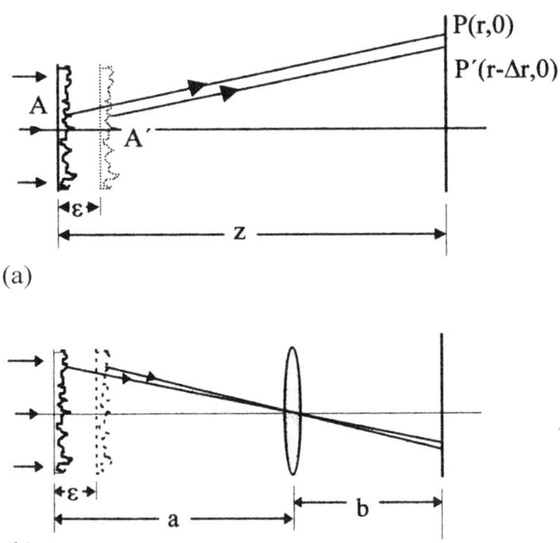

(a)

(b)

Figure 4 Out-of-plane displacement: (a) objective speckle pattern and (b) subjective speckle pattern.

state of speckle will exist at point $P'(r - \Delta r, 0)$; that is, the speckle would have radially shifted by Δr. We therefore have

$$\frac{r}{z} = \frac{r - \Delta r}{z - \varepsilon}$$

or

$$|\Delta r| = \varepsilon \frac{r}{z} \tag{6}$$

A similar situation prevails for a subjective speckle pattern, as shown in Figure 4b. We therefore obtain

$$\frac{a}{a - \varepsilon} = \frac{r + \Delta r}{r}$$

or

$$|\Delta r| = \varepsilon \frac{r}{a} \tag{7}$$

where a is the object distance. Thus, for axial translation of the object, the speckle pattern shifts radially—it either expands or contracts depending on the direction of the shift. It may be noticed that because of the r/z or r/a factor, the radial movement of the speckle pattern requires rather large out-of-plane displacement of the object.

C. Tilt of the Object

Let the object be illuminated in the direction making an angle α with the z axis, which is also in the direction of the surface local normal and the speckle pattern be observed at a point P in the direction making an angle β as shown in Figure 5a. When the object is tilted by a small angle $\Delta\phi$, the speckle pattern shifts to a new position that is angularly shifted by $\Delta\psi$ where $\Delta\psi = (1 + \cos\alpha/\cos\beta)\Delta\phi$. Only for the case of normal illumination and observation directions (angles α and β are very small) is the angular shift of the speckle pattern twice the tilt of the object. The situation is quite different when we consider the subjective

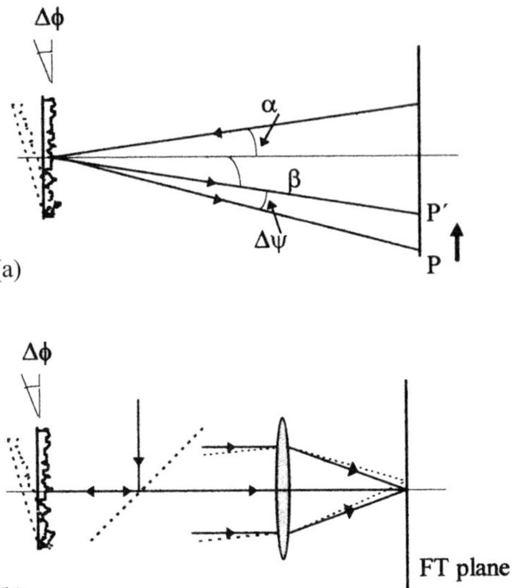

(a) (b)

Figure 5 Motion of the speckle (a) due to tilt of the object (objective speckle pattern) and (b) at the FT plane due to tilt of the object.

speckle pattern. In fact, there is no shift at the image plane due to the tilt of the object. Between the image plane and the focal plane (or FT plane), speckle shift is due to both in-plane and tilt contributions. At the FT plane, the speckle pattern does not shift as a result of in-plane translation but shifts by $\Delta x_f = f\Delta\phi$ when the object is tilted by $\Delta\phi$ (Fig. 5b). When the object is illuminated by a divergent wave, one can find a plane that is sensitive only to the in-plane motion and another plane sensitive to tilt alone.

It is thus seen that a speckle pattern undergoes changes when the object is either translated or tilted. However, for deformation measurement, we are interested in measuring changes and shifts at various points of the objects, hence we employ only the subjective speckle pattern. This way, a correspondence between the object and the image points is maintained. In other words, the local changes in the object cause changes locally in the speckle pattern rather than over the whole speckle pattern plane. The changes in a speckle pattern due to the object deformation are (1) positional changes accompanied by irradiance changes and decorrelation and (2) phase changes that are made visible by adding a reference wave or another speckled wave at the image plane. Often, both these changes occur simultaneously, but one may dominate over the other.

The speckle methods used for deformation measurement and vibration analysis can be placed under the following categories:

1. Speckle photography
2. Speckle interferometry
3. Speckle shear interferometry
4. Electronic speckle pattern interferometery (ESPI) and shear ESPI

Methods 1–3 employ photographic medium for recording and method 4 employs electronic detection. Correlation, as in speckle photography, can also be accomplished digitally.

VI. SPECKLE PHOTOGRAPHY

The object may be illuminated obliquely or normally. We make an image of the object, as shown in Figure 6a, on photographic plate capable of resolving the speckle pattern, and the first exposure is recorded. The object is then loaded and another record of the displaced speckle pattern is made on the same plate. This way we have recorded two speckle patterns, one of them translated locally by **d**. We need to find out **d** at various locations on the plate and then generate the deformation map. It was pointed out earlier that the speckle displacement has poor sensitivity for axial displacements. Hence, speckle photography is used mostly to measure in-plane displacements and in-plane vibration amplitudes.

Let us first examine the specklegram (negative film or plate) realized by making a single exposure. The intensity recorded is given by $I(x, y) = |\mathbf{u}(x, y)|^2$. The amplitude transmittance of this negative (specklegram) is expressed as follows:

$$t(x, y) = t_0 - \beta T I(x, y) \tag{8}$$

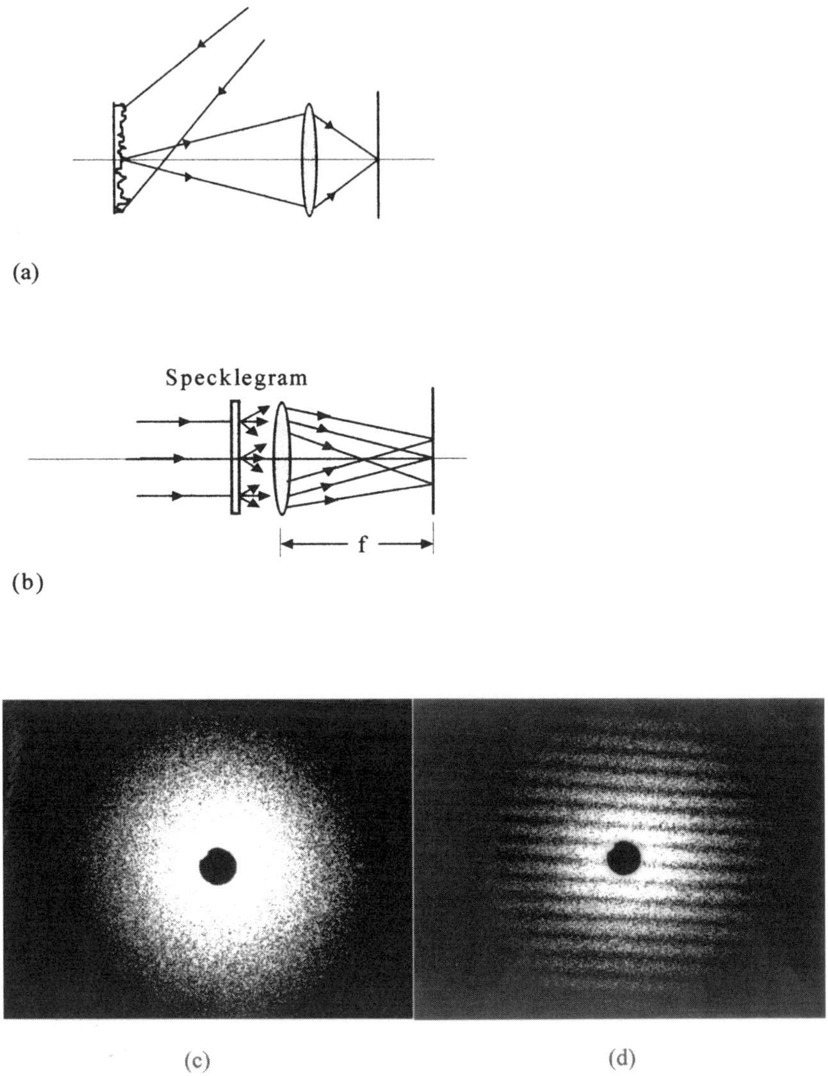

Figure 6 (a) Speckle photography—bodily in-plane displacement. (b) Singly exposed specklegram illuminated by a collimated beam. (c) Halo as formed at the FT plane when a single-exposure specklegram is used in the setup of (b). (d) Halo modulated by fringes when a double-exposure specklegram is placed in setup of (b).

where t_0 is the bias transmittance, β is a constant, and T is the exposure time. Since the speckle pattern consists of a grainy structure, each grain being identified by a δ function, the intensity $I(x, y)$ could also be expressed as follows:

$$I(x, y) = \iint I(x', y')\delta(x - x', y - y')dx'dy' \tag{9}$$

When this specklegram is placed in a setup like that of Figure 6b and illuminated by a parallel beam of light, the amplitude transmitted is given by $\mathbf{u}_0(x, y)t(x, y)$, where $\mathbf{u}_0(x, y)$ is the amplitude of the illuminating plane wave. The specklegram will diffract the light over a reasonably large cone, depending on the speckle size. We collect this diffracted light with a lens and make an observation at the back focal plane of the lens. Essentially, we are taking the Fourier transform of the amplitude transmittance $t(x, y)$ of the speckle record. The amplitude at the FT plane, assuming unit amplitude illumination wave, is given by

$$\begin{aligned} U(\mu, \nu) &= \iint t(x, y)\exp[-2\pi i(\mu x + \nu y)]dx\,dy \\ &= \iint t_0 \exp[-2\pi i(\mu x + \nu y)]dx\,dy \\ &\quad - \beta T \iint I(x, y)\exp[-2\pi i(\mu x + \nu y)]dx\,dy \\ &= t_0\delta(\mu, \nu) - \beta T \Im[I(x, y)] \end{aligned} \tag{10}$$

where $\Im[\ ]$ signifies Fourier transform. The intensity distribution at the FT plane, $|U(\mu, \nu)|^2$, consists of a strong central peak and a light distribution around called the halo. This is shown in Figure 6c; the central portion is blocked while the halo is being recorded. The halo contains a range of spatial frequencies between 0 and $\sigma_s/f\lambda$, where σ_s is the average size in the subjective speckle pattern. The diameter of the halo is $(2f\lambda/\sigma_s)$. The halo distribution is given by the autocorrelation of the aperture function of the imaging lens. A physical insight into the formation of the halo can be obtained if we consider a specklegram as having a large number of sinusoidal gratings of continuously varying pitch and random orientations. When the specklegram is illuminated, these gratings diffract the beam in various directions, forming the halo at the back focal plane of a lens.

Let us now consider a double-exposure specklegram. In the first exposure, an intensity distribution $I_1(x, y)$ is recorded. The object is then deformed and the second exposure $I_2(x, y)$ is recorded on the same plate. The deformation causes the speckle pattern to shift locally. Therefore, the intensity distribution $I_2(x, y)$ can be expressed as follows:

$$I_2(x, y) = \iint I(x', y')\delta(x + d_x - x', y + d_y - y')dx'dy' \tag{11}$$

where d_x and d_y are the components of $\mathbf{d}$ along the x and y directions, respectively. The total intensity recorded is

$$\begin{aligned} I_t(x, y) &= I_1(x, y) + I_2(x, y) \\ &= \iint I(x', y')\{\delta(x - x', y - y') \\ &\quad + \delta(x + d_x - x', y + d_y - y')\}dx'dy' \end{aligned} \tag{12}$$

Again, if this double-exposure specklegram is illuminated by a collimated beam, one obtains, at the focal plane of the lens, a central order and the superposition of halos belonging to the initial and final states of the object. Mathematically, the amplitude transmittance of the double-exposure specklegram is

$$t(x, y) = t_0 - \beta T\{I_1(x, y) + I_2(x, y)\} \tag{13}$$

The amplitude at the FT plane is given by

$$\Im[t(x, y)] = t_0\delta(\mu, \nu) - \beta T\Im[I_1(x, y) + I_2(x, y)] \tag{14}$$

For simplicity, we now confine ourselves to one dimension, hence write the total intensity as follows:

$$I_t(x) = \iint I(x')\{\delta(x - x') + \delta(x + d_x - x')\}dx'dy' \tag{15}$$

which is a convolution of $I(x)$ with $[\delta(x) + \delta(x + d_x)]$. Therefore

$$\Im[I_t(x)] = \Im[I_1(x, y) + I_2(x, y)] = \Im[I(x)]\Im[\delta(x) + \delta(x + d_x)] \tag{16}$$

The Fourier transform of $\delta(x)$ is a plane wave propagating on-axis, while that of $\delta(x + d_x)$ is also a plane wave propagating inclined with the axis: that is,

$$\begin{aligned}\Im[\delta(x) + \delta(x + d_x)] &= \int \delta(x)\exp(-2\pi i\mu x)dx \\ &\quad + \int \delta(x + d_x)\exp(-2\pi i\mu x)dx \\ &= c + c\exp(2\pi i\mu d_x)\end{aligned} \tag{17}$$

where c is a constant, being the Fourier transform of a δ function. The spatial frequency μ is defined as $\mu = x_f/f\lambda$, with x_f being the x coordinate on the FT plane and f the focal length of the lens. Suppressing the central order, we can write for the intensity distribution in the FT plane (in the halo)

$$I(\mu) \propto 2\beta^2T^2c^2(1 + \cos 2\pi\mu d_x) = I_0(\mu)\cos^2 \pi\mu d_x \tag{18}$$

It is therefore seen that the halo, given by $I_0(\mu)$, is modulated by the $\cos^2(\pi\mu d_x)$ term. In other words, a fringe pattern appears in the halo. Figure 6d shows the fringe pattern in the halo. This fringe pattern is similar to the Young's double-aperture fringes. The fringes are termed Young's fringes. In the $\cos^2(\pi\mu d_x)$ term, there are two variables, namely μ, the coordinate along the x axis, and d_x, which is a function of local coordinates if the deformation is not constant over the whole specklegram. For constant d_x, the interpretation is simple and straightforward: the halo distribution is modulated by the cosinusoidal fringes whose spacing is $\Delta x_f = f\lambda/d_x$. Therefore, the magnitude of d_x may be obtained from the fringe width measurement. However, if d_x is not constant, each value of d_x will produce its own fringe pattern, which on superposition may completely wash out this intensity variation. In such a situation, the specklegram is interrogated with a narrow or unexpanded laser beam to extract the displacement information from each region of interrogation. The displacement in the region of interrogation should be constant.

VII. METHODS OF EVALUATION

A double-exposure specklegram contains two shifted speckle patterns, and to generate the deformation map, the shift must be determined at a number of locations on the specklegram. It was earlier remarked that the Young's type of fringes are formed when the specklegram is illuminated by a narrow beam. These fringes give the direction and the magnitude of the displacement at a point. The process of extracting the information from a specklegram is called filtering. The filtering is done both at the plane of the specklegram and also at its Fourier transform plane. These methods are called as pointwise filtering and wholefield filtering.

One more method of filtering, usually applicable to out-of-plane displacement measurement, is known as Fourier filtering. It can also be used in other cases; the fringes are generally localized on the specklegram.

A. Pointwise Filtering

It was mentioned earlier that if the displacement of the speckles is nonuniform, a fringe pattern may not be formed at the Fourier plane when the whole specklegram is illuminated. However, it is safe to assume that the speckle movement over a very small region of the image (specklegram) is uniform. Therefore, if the double-exposure specklegram is illuminated by a narrow (unexpanded) beam and the observation is made at a plane sufficiently far away (far field) as shown in Figure 7a, a system of Young's fringes will be formed. The fringes always run perpendicular to the direction of displacement and the fringe width $\bar{p}$ is inversely proportional to the displacement (i.e., $\bar{p} = \lambda z/|\mathbf{d}|$). Thus, both the direction and the magnitude of the displacement at each interrogation region on the specklegram are obtained. The sign ambiguity is

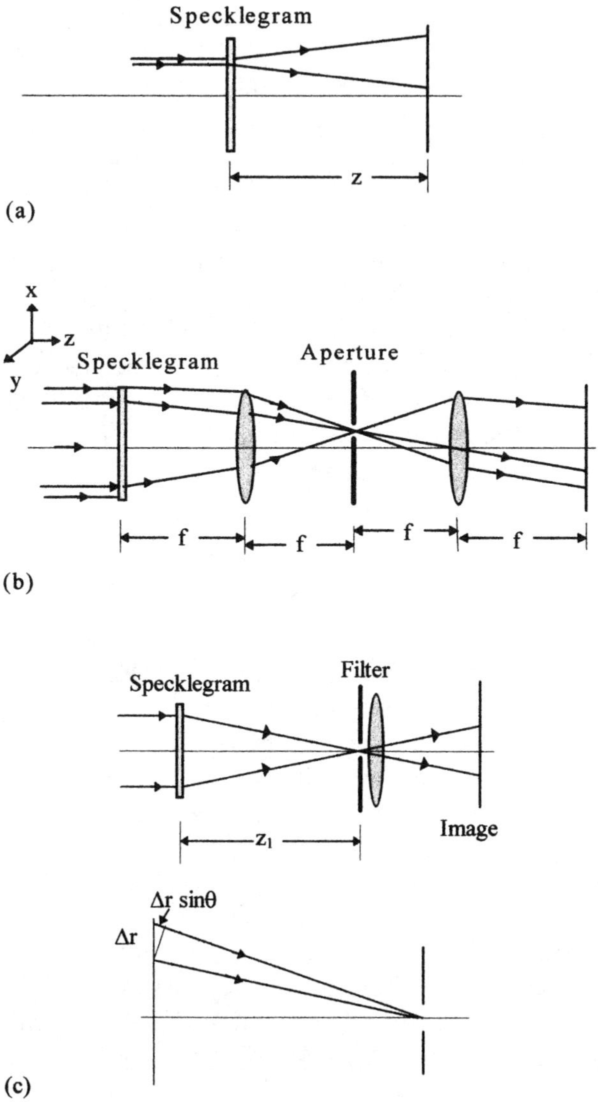

Figure 7 (a) Pointwise filtering arrangement. (b) Wholefield filtering arrangement. (c) Fourier filtering arrangement.

still not resolved. The sign ambiguity can be resolved by giving a linear known displacement to the photographic plate before the second exposure. When the magnitude and the direction of displacement are obtained at a large number of points on the specklegram, the displacement map on the specklegram is generated, which is then translated to the object surface through the magnification of the imaging system.

The contrast of the Young's fringes is influenced by several factors, the most important ones being nonuniform displacement and missing speckle pairs in the region of illumination. Further, it may be kept in mind that the halo distribution is nonuniform and because of background intensity, the Young's fringes may not have unit contrast. Therefore, the maxima and minima of the fringes are shifted. This shift introduces errors in the calculation of the displacement. Methods have been developed to correct for this.

B. Wholefield Filtering

We consider the arrangement shown in Figure 7b. This is a $4f$ configuration. The filtering is carried out at the FT plane by an aperture. The image of the specklegram is formed at unit magnification on the output plane by the light allowed through the filtering aperture. The image contains the fringes, which represent constant in-plane displacements in the direction of the filtering aperture. To understand the working of the wholefield filtering technique, let us place an aperture of adequate size at a position x_{f} along the x direction. All the identical pair scatterers on the specklegram will diffract light in phase in the direction of the filtering aperture if their separation is such that the waves diffracted by these point scatterers have a path difference of integral multiples of λ, that is,

$$d_x \sin\theta = m\lambda \qquad \text{for } m = 0, \pm 1, \pm 2, \pm 3, \ldots \tag{19}$$

where d_x is the x component of the displacement vector and $\sin\theta = x_{\mathrm{f}}/f$. Thus, we obtain

$$d_x = \frac{m\lambda f}{x_\mathrm{f}} \tag{20a}$$

These areas therefore appear bright in the image, forming bright fringes that are loci of constant d_x. Similarly, when the filtering aperture is placed at y_f along the y axis, we obtain d_y as

$$d_y = \frac{m'\lambda f}{y_\mathrm{f}} \tag{20b}$$

These fringes are loci of constant d_y. In other words, the fringes represent the contours of constant in-plane displacement components that are separated by incremental displacements Δd_x and Δd_y, where

$$\Delta d_x = \frac{\lambda f}{x_\mathrm{f}} \tag{21a}$$

and

$$\Delta d_y = \frac{\lambda f}{y_\mathrm{f}} \tag{21b}$$

This is a wholefield method: the fringes are formed over the whole object surface. To obtain d_x and d_y, we need to know m and m', hence any regions on the object where no displacement has taken place. It is also seen that the sensitivity of the method is variable, being maximum when the filtering aperture is placed at the periphery of the diffraction halo; the increase in sensitivity follows the decrease in the available light for image formation.

C. Fourier Filtering: Measurement of Out-of-Plane Displacement

It has been shown that longitudinal or axial displacement ε of the object results in a radial displacement of the speckle pattern (i.e., $\Delta r = \varepsilon r/z$). The displacement magnitude ε is obtained by Fourier filtering. The specklegram is illuminated by a parallel beam as shown in Figure 7c, and an aperture is placed on the optical axis for filtering. A lens placed just behind the aperture images the specklegram at the observation plane. All those point pairs (identical scatterers) will diffract the incident light in phase at the filtering plane if $\Delta r \sin\theta = m\lambda$, where $\sin\theta = r/z_1$ and λ is the wavelength of light used for filtering (Fig. 7c). Thus

$$\Delta r \sin\theta = \frac{\varepsilon r}{z}\frac{r}{z_1} = m\lambda$$

or

$$\frac{\varepsilon r^2}{z z_1} = m\lambda \tag{22}$$

This indicates that a circular fringe pattern is observed at the observation plane. The displacement ε is obtained by measuring the radii of fringes of different orders as follows:

$$\varepsilon = \frac{n\lambda z z_1}{(r_{m+n}^2 - r_m^2)} \tag{23}$$

where r_{m+n} and r_m are the radii of the $(m+n)$th and mth-order circular fringe, respectively.

VIII. SPECKLE PHOTOGRAPHY WITH VIBRATING OBJECTS: IN-PLANE VIBRATION

Let us consider an object that is executing an in-plane vibration of amplitude $A(x, y)$. The object is imaged on the photographic plate. As a result of in-plane vibration, a speckle stretches to a line of length $2A(x, y)m$, where m is the magnification. As mentioned earlier, a sinusoidally vibrating object spends more time at the positions of maximum displacements. Therefore, the speckle line has a nonuniform brightness distribution. However, if the recording is done on a high contrast photographic plate, the speckle line would have a uniform density after development. We therefore assume that the speckles are uniformly elongated to $2A(x, y)m$. When such a specklegram (time-averaged recording) is pointwise-filtered, the halo distribution is modulated by $\text{sinc}^2(2\mu m A)$. The zeros of this function occur wherever

$$2\mu m A = n \qquad \text{for } n = 0, \pm 1, \pm 2, \pm 3, \ldots$$

or

$$A(x, y) = \frac{n\lambda z}{2m x_{\mathrm{f}_n}} \tag{24}$$

where x_{f_n} is the position of the nth fringe. From this expression, the amplitude of vibration at any location on the specklegram can be found.

In fact, when an object vibrates, the speckles on the regions that are moving smear out and the contrast of the speckles is reduced. The nodal regions where there is no movement have a unit contrast. Therefore, one observes a low contrast vibration mode pattern with the nodal lines appearing dark. A visual speckle interferometer was proposed by Burch and later modified by Stetson. This interferometer uses a reference beam to visualize out-of-plane vibration amplitudes.

IX. SENSITIVITY OF SPECKLE PHOTOGRAPHY

The halo size is governed by the speckle size which, in turn, is governed by the F# of the imaging lens. For the speckle displacement to be measurable, at least one fringe should be formed within a halo. In other words, the fringe width must be equal to or less than the halo diameter. Indeed, one fringe will be formed within the halo when the speckle shifts by an amount equal to its average size. This probably is the lower limit of the displacement that speckle photography can sense. The upper limit is also set by the speckle size. When the fringe width becomes equal to the speckle size, the fringes will not be discernible. In fact, to ensure that the fringe will be discernible, we choose a fringe width of approximately 10 times the speckle size. This therefore sets an upper limit for speckle displacement measurement. For wholefield filtering, similar limits apply. The quality of fringes in wholefield filtering is considerably poorer than in pointwise filtering.

One of the serious drawbacks of speckle photography is that nonuniform halo distribution results in a shift in the positions of the maxima of the fringes in the fringe pattern. The same is true for the minima positions when some background is present. Errors are introduced by this effect on the measured values of the displacements. Methods are available to correct for these errors.

X. PARTICLE IMAGE VELOCIMETRY

Speckle photography of seeded flows carried out with pulsed lasers or scanned laser beams provides information about the flow (spatial variation of the velocity of flow—in fact the velocity of seeds). The technique is known as particle image velocimetry (PIV). First, a record is made with a short duration laser pulse so that the motion is frozen, and then the second exposure is made a short time later. The seeds (particles) would have moved during this time to new locations depending on their velocities. The double-exposure record is pointwise filtered to generate the velocity map.

XI. WHITE LIGHT SPECKLE PHOTOGRAPHY

Most objects are good candidates for white light speckle photography, since their surface structure is quite adequate for this. However, this property is enhanced by coating the surface with a retroreflective paint. The images of the embedded glass balls in retroreflective paint act as speckles. Deformation studies, based on principles similar to those applied to laser speckles, can therefore be done by means of white light speckles.

XII. SHEAR SPECKLE PHOTOGRAPHY

It is obvious that speckle photography is a tool to essentially measure in-plane displacement or the in-plane component of deformation because it has very low sensitivity to the out-of-plane component. It is also used to measure bodily tilts. A stress analyst is often interested in strain rather than displacement. Strain is obtained from displacement data by numerical differentiation, a procedure which is error-prone. However, strain can be obtained by optical differentiation, that is, by obtaining displacement values at two points that are separated but closely spaced. With a double-exposure specklegram, this can be achieved by interrogating it by two parallel-displaced narrow beams. Each beam produces Young's-type fringes in the diffraction halo, and the superposition of the two patterns (i.e., from each illumination beam), will generate a moiré pattern, from which strain can be calculated (Chapter 8). Alternatively, two states of the object can be recorded on two separate films or plates. During filtering, one plate is displaced with respect to the other to introduce shear. Again, a moiré pattern is formed from which strain is calculated. Both these techniques provide a variable shear by varying the beam separation or specklegram displacement.

When two narrow beams are used for illumination, the diffraction halos are spatially displaced. For low angle diffraction, the overlap region may be very small to provide a meaningful moiré pattern. On the other hand, when two double-exposure films or plates are used, the underlying assumption that the object has been identically loaded for each double-exposure record may be difficult to realize. It is therefore desirable to record a single shear double-exposure specklegram. For this purpose, we use a pair of wedge plates in front of the imaging lens: each wedge plate carries a sheet polarizer. The transmission axes of the polarizers are crossed. The object is illuminated either with a circularly polarized light or a linearly polarized light having its azimuth at 45° with the polarizer axis. A double-exposure record, with object loaded in between, can be recorded. The interrogation of such a record by a narrow beam will generate a moiré pattern. The drawback of this method is that it has a fixed shear determined by the wedge angles. In addition, the Young's fringes have a nonuniform intensity, are noisy, and are few in number, hence moiré fringes are of poor contrast. Figure 8 shows a moiré pattern obtained from such an arrangement.

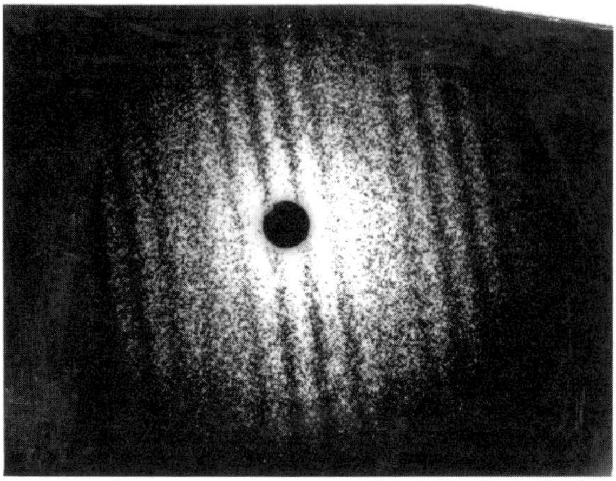

Figure 8 Moiré pattern due to the superposition of two Young's fringe patterns.

XIII. SPECKLE INTERFEROMETRY

Although the speckle phenomenon is itself essentially an interference phenomenon, when a reference beam is added to the speckle pattern to code its phase, the technique is then termed as speckle interferometry. Speckle interferometry was first applied to measure in-plane displacements by Leendertz. The sensitivity of the measurement could be increased over that of speckle photography, which was limited by the speckle size. The basic theory was borrowed from holographic interferometry, since the phase difference introduced by deformation is governed by the same equation, namely, $\delta = (\mathbf{k}_2 - \mathbf{k}_1) \cdot \mathbf{d}$. When the the object is illuminated with two beams with directions symmetrical to the object normal (also the optical axis) and observation is made along the optical axis, the arrangement generates fringes that are contours of constant in-plane displacement. For reasons that will become obvious in later sections, these fringes are called correlation fringes.

The schematic of the setup for measuring in-plane displacement is shown in Figure 9. The object is illuminated by two plane waves incident symmetrically at angles θ and $-\theta$ with respect to the optical axis. An image of the object is made on the recording

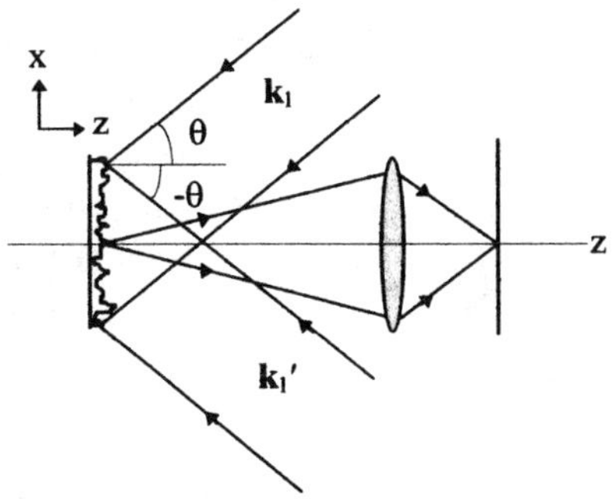

Figure 9 In-plane displacement measurement by speckle interferometry.

material—say, a photographic plate. The object is deformed, and a second exposure is made on the same plate. On filtering, this double-exposure record generates fringes representing in-plane component along the x direction. The whole process can be explained mathematically as follows.

Let us consider that the object is illuminated by unit amplitude plane waves, which are represented as $\exp(ikx\sin\theta)$ and $\exp(-ikx\sin\theta)$ at the plane of the object. The net amplitude of the waves at the plane of the object is $[\exp(ikx\sin\theta) + \exp(-ikx\sin\theta)]$. Let the scattering process from the object be represented by $\Re(x, y)$, which is a complex function and includes the surface characteristics. Thus, the field on the object surface just after scattering is

$$\Re(x, y)[\exp(ikx\sin\theta) + \exp(-ikx\sin\theta)]$$

The amplitude at the image plane is obtained through the superposition integral, that is,

$$\begin{aligned} u(x_i, y_i) &= \iint \Re(x, y)[\exp(ikx\sin\theta) + \exp(-ikx\sin\theta)] \\ &\quad h(x_i - mx, y_i - my)dx\,dy \\ &= a_1(x_i, y_i)\exp[i\phi_1(x_i, y_i)] + a_2(x_i, y_i)\exp[i\phi_2(x_i, y_i)] \end{aligned} \tag{25}$$

where $h(x, y)$ is the impulse function and m is the magnification of the imaging lens; $a_1(x_i, y_i)$, $\phi_1(x_i, y_i)$, $a_2(x_i, y_i)$, $\phi_2(x_i, y_i)$ are the amplitudes and phases of the speckled waves at the image plane due to each of the illuminating waves. When the object is displaced by d_x in the x direction, the amplitude in the image may be written as follows:

$$\begin{aligned} u'(x_i, y_i) &= \iint \Re(x + d_x, y)\{\exp[ik(x + d_x)\sin\theta] \\ &\quad + \exp[-ik(x + d_x)\sin\theta]\}h(x_i - mx, y_i - my)dx\,dy \\ &= a_1(x_i, y_i)\exp[i\phi_1(x_i, y_i)]\exp(ikd_x\sin\theta) \\ &\quad + a_2(x_i, y_i)\exp[i\phi_2(x_i, y_i)]\exp(-ikd_x\sin\theta) \end{aligned} \tag{26}$$

This equation was arrived at by a tacit assumption that $\Re(x + d_x, y) = \Re(x, y)$. This assumption implies that the speckle displacement at the object surface is not attributable to the motion and that the surface characteristics do not change over a distance d_x. In fact, there is a displacement of speckle, but we assume that the speckle displacement is much smaller than the speckle size.

Now, we can write the intensities recorded in both the exposures as follows:

$$\begin{aligned} I_1(x_i, y_i) &= a_1^2 + a_2^2 + 2a_1a_2\cos(\phi_1 - \phi_2) \\ &= a_1^2 + a_2^2 + 2a_1a_2\cos\phi \end{aligned} \tag{27a}$$

$$I_2(x_i, y_i) = a_1^2 + a_2^2 + 2a_1a_2\cos(\phi + 2kd_x\sin\theta) \tag{27b}$$

For subsequent analyses of the various techniques, we will write the intensity distribution in the second exposure as follows:

$$I_2(x_i, y_i) = a_1^2 + a_2^2 + 2a_1a_2\cos(\phi + \delta) \tag{28}$$

where $\delta = 2kd_x\sin\theta = [(4\pi/\lambda)d_x]\sin\theta$. Later, we will use different arguments to prove that δ is indeed equal to $2kd_x\sin\theta$ for the experimental configuration of Figure 9. For the present, we mention that ϕ_1, ϕ_2, a_1, a_2 are random variables becuase $\Re(x, y)$ is a random variable. Therefore, $\phi = (\phi_1 - \phi_2)$ is also a random variable. The total intensity recorded will be

$$I_1 + I_2 = a_1^2 + a_2^2 + 4a_1a_2 \cos\left(\phi + \frac{\delta}{2}\right) \cos\frac{\delta}{2} \tag{29}$$

In fact, examination of Eqs. (27a) and (28) for I_1 and I_2 reveals that when $\delta = 2m\pi$, the equations are identical—the speckles in both exposures are fully correlated. When $\delta = (2m + 1)\pi$, the speckles are uncorrelated. This correlation does provide the basis for fringe formation.

It should be noted that the contrast of the fringes is very poor in this arrangement, mainly because of the very strong granular bias term $(a_1^2 + a_2^2)$. We will discuss other arrangements in which this bias term has been isolated, thereby improving the contrast of the fringes. First, let us prove that the phase change δ due to in-plane displacement d_x is indeed $2kd_x \sin\theta$. The phase change δ due to in-plane displacement can be expressed as follows:

$$\delta = (\delta_2 - \delta_1) = (\mathbf{k}_2 - \mathbf{k}_1) \cdot \mathbf{d} - (\mathbf{k}_2 - \mathbf{k}'_1) \cdot \mathbf{d} = (\mathbf{k}'_1 - \mathbf{k}_1) \cdot \mathbf{d} \tag{30}$$

where $\mathbf{d} = (d_x\bar{i} + d_y\bar{j} + d_z\bar{k})$, and $\mathbf{k}'_1$ and $\mathbf{k}_1$ are the propagation vectors of the illumination beams. The equation of phase difference $\delta_2 = (\mathbf{k}_2 - \mathbf{k}_1) \cdot \mathbf{d}$ was derived in Chapter 5 and is applicable to speckle interferometry. From the geometry of the experimental configuration of Figure 9, $\mathbf{k}'_1$ and $\mathbf{k}_1$ are expressed as follows:

$$\mathbf{k}'_1 = \frac{2\pi}{\lambda}(\sin\theta\bar{i} - \cos\theta\bar{k}) \tag{31a}$$

and

$$\mathbf{k}_1 = \frac{2\pi}{\lambda}(-\sin\theta\bar{i} - \cos\theta\bar{k}) \tag{31b}$$

Thus the vector $(\mathbf{k}'_1 - \mathbf{k}_1)[= (2\pi/\lambda)2\sin\theta\bar{i}]$ lies in the plane of the object. Hence

$$\delta = (\mathbf{k}'_1 - \mathbf{k}_1) \cdot \mathbf{d} = \left(\frac{2\pi}{\lambda} 2 \sin\theta \bar{i}\right)(d_x \bar{i} + d_y \bar{j} + d_z \bar{k}) = \frac{2\pi}{\lambda} 2 d_x \sin\theta$$

Bright fringes are formed when

$$\frac{2\pi}{\lambda} 2 d_x \sin\theta = 2m\pi$$

Therefore

$$d_x = \frac{m\lambda}{2\sin\theta} \tag{32}$$

The arrangement is sensitive only to the in-plane component of displacement. Consecutive in-plane fringes differ by $\lambda/(2\sin\theta)$. Obviously, by varying the interbeam angle, one can vary the sensitivity over a very wide range from zero to $\lambda/2$ per fringe. Furthermore, this particular arrangement has the following features:

As seen from the theoretical analysis, the contribution of the out-of-plane displacement component d_z has been fully compensated for collimated illumination. This is because both the waves suffer equal phase changes arising from out-of-plane displacement; hence the net phase change is zero.

The arrangement is not sensitive to the y component (i.e., d_y). However, d_y can be measured by rotating the experimental arrangement through 90°.

It offers variable sensitivity.

The disadvantage is that the fringes obtained are of low contrast.

XIV. THE CORRELATION COEFFICIENT IN SPECKLE INTERFEROMETRY

The correlation coefficient of two random variables X and Y is defined by

$$\rho_{XY} = \frac{\langle XY \rangle - \langle X \rangle \langle Y \rangle}{\sigma_X \sigma_Y} \tag{33}$$

where $\sigma_X^2 = \langle X^2 \rangle - \langle X \rangle^2$ and $\sigma_Y^2 = \langle Y^2 \rangle - \langle Y \rangle^2$ and $\langle\ \rangle$ stands for ensemble average. If X and Y are uncorrelated, then $\langle XY \rangle = \langle X \rangle \langle Y \rangle$ and the correlation coefficient is zero as expected. The correlation coefficient of the random variables $I_1(x_i, y_i)$ and $I_2(x_i, y_i)$, the intensities recorded in the first and second exposures, is given by

$$\rho(\delta) = \frac{\langle I_1 I_2 \rangle - \langle I_1 \rangle \langle I_2 \rangle}{[\langle I_1^2 \rangle - \langle I_1 \rangle^2]^{1/2} [\langle I_2^2 \rangle - \langle I_2 \rangle^2]^{1/2}} \tag{34}$$

where the intensities I_1 and I_2 are expressed as follows:

$$I_1(x, y) = a_1^2 + a_2^2 + 2a_1 a_2 \cos\phi = i_1 + i_2 + 2\sqrt{i_1 i_2} \cos\phi \tag{35a}$$

and

$$I_2(x, y) = i_1 + i_2 + 2\sqrt{i_1 i_2} \cos(\phi + \delta) \tag{35b}$$

In this expression, the intensities i_1, and i_2 refer to the intensities at the image plane due to individual illuminating beams. Equation (34) for the correlation coefficient is evaluated by noting the following

The intensities i_1, i_2, and the phase ϕ are independent random variables, hence can be averaged separately.

$\langle \cos\phi \rangle = \langle \cos(\phi + \delta) \rangle = 0$

$\langle i_1^2 \rangle = 2\langle i_1 \rangle^2$

When the intensity values I_1 and I_2 are substituted in Eq. (34) for correlation coefficient and the averages taken, we obtain

$$\rho(\delta) = \frac{\langle i_1^2 \rangle + \langle i_2^2 \rangle + 2\langle i_1 \rangle \langle i_2 \rangle \cos\delta}{(\langle i_1 \rangle + \langle i_2 \rangle)^2} \tag{36}$$

The correlation coefficient depends on the intensities of the beams and the phase introduced by deformation. If we consider that $\langle i_1 \rangle = r\langle i_2 \rangle$ (i.e., one beam is r times stronger than the other beam), the correlation coefficient takes the following simpler form:

$$\rho(\delta) = \frac{1 + r^2 + 2r\cos\delta}{(1 + r)^2} \tag{37}$$

This has a maximum value of unity when $\delta = 2m\pi$ and minimum value of $(1 - r)^2/(1 + r)^2$ when $\delta = (2m + 1)\pi$. The minimum value will be zero if $r = 1$ (i.e., when the average intensities of both beams are equal). The correlation coefficient then would vary between 0 and 1 as the value of δ varies over the record. This situation is completely at variance with holographic interferometry, where the reference beam is taken as being stronger than the object beam.

XV. OUT-OF-PLANE SPECKLE INTERFEROMETER

An interferometer for measuring out-of-plane displacement is shown in Figure 10. It is the Michelson interferometer in which one of the mirrors has been replaced by an object under study. This therefore has a reference wave that is smooth or specular. The lens L_2 makes an image of the object at the recording plane. The record consists of an interference between the smooth reference wave and the speckle field in the image of the object. The second exposure is

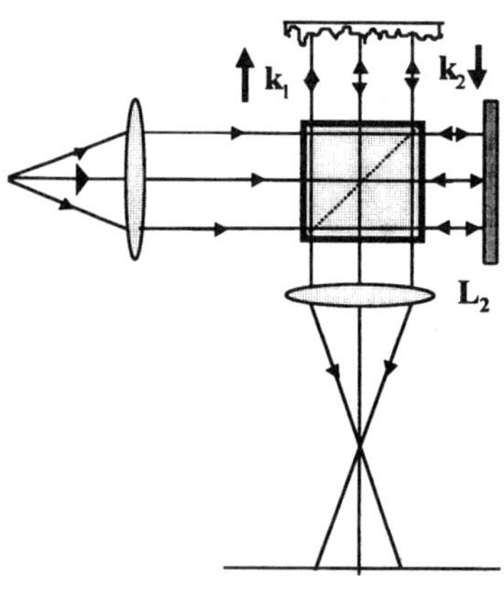

Figure 10 Out-of-plane displacement measurement by speckle interferometry.

recorded on the same plate after the object is loaded. This double-exposure specklegram (interferogram), when filtered, yields fringes that are contours of constant out-of-plane displacement. As before, the intensity distribution in the first exposure is

$$I_1(x_i, y_i) = a_1^2 + r_0^2 + 2a_1 r_0 \cos(\phi_1 - \phi_r) = a_1^2 + r_0^2 + 2a_1 r_0 \cos\phi \tag{38}$$

where a_1, ϕ_1, and ϕ are random variables. The second exposure records the intensity distribution given by

$$I_2(x_i, y_i) = a_1^2 + r_0^2 + 2a_1 r_0 \cos(\phi + \delta) \tag{39}$$

where the phase difference δ, introduced by deformation, is given as follows:

$$\delta = (\mathbf{k}_2 - \mathbf{k}_1) \cdot \mathbf{d} = \frac{2\pi}{\lambda} 2d_z$$

The phase difference depends only on the out-of-plane component d_z. Bright fringes are formed wherever

$$\frac{2\pi}{\lambda} 2d_z = 2m\pi$$

or

$$d_z = \frac{m\lambda}{2} \tag{40}$$

Thus, consecutive fringes are separated by out-of-plane displacements of $\lambda/2$. Again, the fringes are of low contrast. However, it may be noted that because of imaging geometry and customized configurations, only one of the components of deformation is sensed, unlike the case in HI, where the phase difference δ introduced by deformation invariably depends on all the three components. Further, the correlation fringes are localized at the plane of the specklegram, whereas in HI the fringe pattern is localized in space in general.

XVI. IN-PLANE MEASUREMENT: DUFFY'S METHOD

In the method due to Leendertz described earlier, the object is illuminated symmetrically with respect to the surface normal and observation is made along the bisector of the angle enclosed by the illuminating beams. This method, which is also called a dual-illumination, single-observation-direction method, has very high sensitivity but poor contrast fringes.

It is also possible to illuminate the object along one direction and make an observation along two different directions symmetrical to the optical axis that is also along the local normal to the object. This method is due to Duffy and was developed in the context of moiré gauging. The method is also known as single-illumination, dual-observation-direction method. Figure 11a

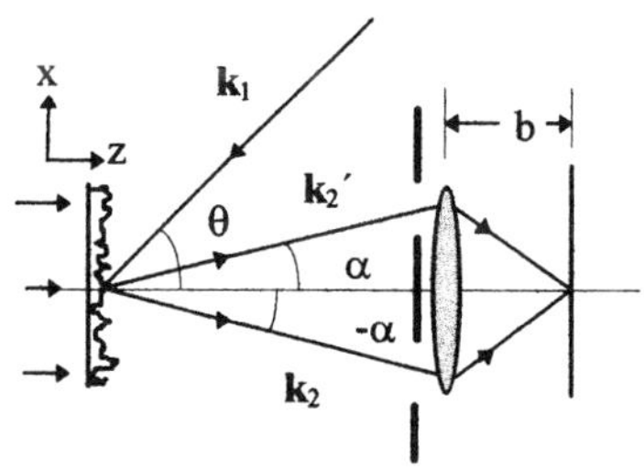

(a)

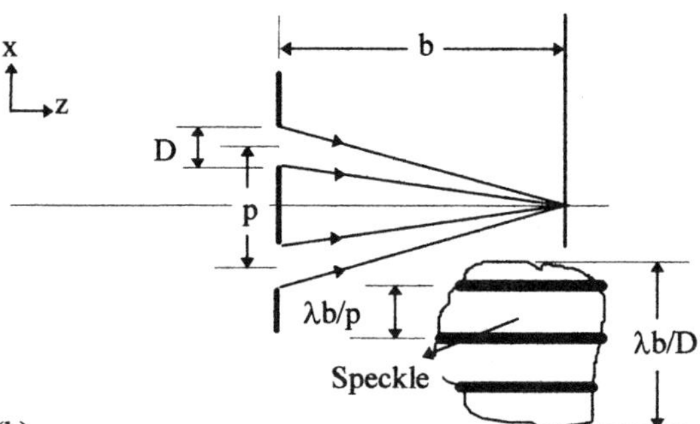

(b)

Figure 11 (a) Duffy's two-aperture arrangement. (b) Modulation of speckle in Duffy's arrangement.

shows a schematic of the experimental arrangement. The object is illuminated at an angle θ, and a two-aperture mask is placed in front of the lens. The apertures enclose an angle of 2α at the object distance. The lens makes an image of the object via each aperture—these images are perfectly superposed.

Each wave passing through the aperture generates a speckled image with a speckle size $\lambda b/D$, where D is the aperture size. These waves are obliquely superposed; hence each speckle is modulated by a fringe pattern when recorded. The fringe spacing in the speckle is $\lambda b/p$, where p is the interseparation between the two apertures, as shown in Figure 11b. When the object is deformed, these fringes shift in the speckle. When the deformation is such that a fringe moves by one period or a multiple thereof, it is then exactly superposed on the earlier recorded position: the fringe contrast is high. The region will diffract strongly on filtering; hence such areas will appear bright.

On the other hand, if the deformation is such that the fringe pattern moves by half a fringe width or odd multiple of half-period, the new pattern will fall midway of the earlier recorded pattern, resulting in almost complete washout of the fringe pattern. These regions will not diffract, or will diffract poorly and appear dark on filtering. Therefore, bright and dark fringes correspond to regions in which the displacement is an integral multiple of $\lambda b/p[= \lambda/(2\sin\alpha)]$ and an odd-integral multiple of $0.5(\lambda b/p)$, respectively.

Mathematically, we write the amplitudes of the waves via each aperture as follows:

$$a_{11} = a_0 \exp(i\phi_{11}) \tag{41a}$$

$$a_{12} = a_0 \exp[i(\phi_{12} + 2\pi\beta x)] \tag{41b}$$

where $\beta(= p/\lambda b)$ is the spatial frequency of the fringe pattern in the speckle pattern. The intensity recorded in the first exposure is

$$I_1(x, y) = |a_{11} + a_{12}|^2 = a_0^2 + a_0^2 + 2a_0^2 \cos(\phi_1 + 2\pi\beta x) \quad \text{for } \phi_1 = \phi_{12} - \phi_{11} \tag{42}$$

Similarly, when the object is loaded, the waves acquire additional phase changes δ_1 and δ_2, respectively; that is, they can be expressed as follows:

$$a_{21} = a_0 \exp[i(\phi_{11} + \delta_1)] \tag{43a}$$

$$a_{22} = a_0 \exp[i(\phi_{12} + 2\pi\beta x + \delta_2)] \tag{43b}$$

The intensity distribution recorded in the second exposure is then

$$I_2(x, y) = |a_{21} + a_{22}|^2 = a_0^2 + a_0^2 + 2a_0^2 \cos(\phi_1 + 2\pi\beta x + \delta) \tag{44}$$

where $\delta(= \delta_2 - \delta_1)$ is the phase difference introduced by the deformation. The specklegram is now ascribed a transmittance $t(x, y)$, where

$$t(x, y) = t_0 - \beta T(I_1 + I_2) \tag{45}$$

where β is a constant and T the exposure time. Information about the deformation is extracted by filtering.

XVII. FILTERING

The specklegram is placed in the setup shown in Figure 12 and is illuminated by a collimated beam, say, a unit amplitude wave. The

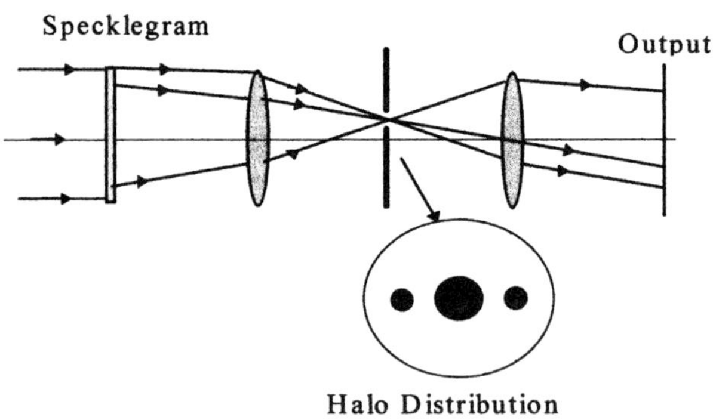

Figure 12 Wholefield filtering of a specklegram recording with Duffy's configuration.

field at the FT plane is $\Im[t(x, y)]$. This consists of a halo distribution with a very strong central peak, as shown in Figure 12. As a result of the grating like structure in each speckle, the halo distribution is modified: it has a central halo (zeroth order) and $\pm$ first-order halos. The zeroth order arises as a consequence of the terms $a_0^2 + a_0^2$, which do not carry any information. Since the speckle size is now larger because smaller apertures were used for imaging, the halo size (zeroth order halo) shrinks. The filtering is done by choosing any one of the first-order halos. A fringe pattern of almost unit contrast is obtained, since the halo (zeroth order) that carries no information has been isolated.

A. Fringe Formation

The phase differences δ_2 and δ_1, due to the deformation, experienced by the waves passing through the apertures can be expressed as follows:

$$\delta_2 = (\mathbf{k}_2' - \mathbf{k}_1) \cdot \mathbf{d}$$
$$\delta_1 = (\mathbf{k}_2 - \mathbf{k}_1) \cdot \mathbf{d}$$

Hence the phase difference $\delta(= \delta_2 - \delta_1)$ is given by

$$\delta = (\mathbf{k}_2' - \mathbf{k}_2) \cdot \mathbf{d} \tag{46}$$

Since the illumination beam and observation beams lie in the x–z plane, the wave vectors $\mathbf{k}_2'$ and $\mathbf{k}_2$ are expressed as follows:

$$\mathbf{k}_2' = \frac{2\pi}{\lambda}(\sin\alpha\bar{i} + \cos\alpha\bar{k})$$

and

$$\mathbf{k}_2 = \frac{2\pi}{\lambda}(-\sin\alpha\bar{i} + \cos\alpha\bar{k})$$

Thus

$$\delta = (\mathbf{k}_2' - \mathbf{k}_2) \cdot \mathbf{d} = \frac{2\pi}{\lambda} 2d_x \sin\alpha$$

Bright fringes are formed wherever

$$\frac{2\pi}{\lambda} 2d_x \sin\alpha = 2m\pi$$

or

$$d_x = \frac{m\lambda}{2\sin\alpha} \tag{47}$$

This result is similar to that obtained earlier for the Leendertz method except that the angle θ is replaced by the angle α. Obviously, α cannot take very large values—the magnitude of α is determined by the lens aperture or by F#. Therefore, the method has intrinsically poor sensitivity. At the same time, the speckle size is very large compared to that in the Leendertz method; hence the range of in-plane displacement measurement is large. The method yields high contrast fringes because the unwanted speckled field has been removed by the grating like structure formed during recording.

It is indeed very simple and easy to extend Duffy's method to measure both the components of in-plane displacement simultaneously. We describe here two configurations. One uses the aperture configuration shown in Figure 13a. The apertures are located at the four corners of a square. In another configuration (Fig. 13b), only three apertures are used: these apertures lie on the vertices of a right-angled triangle. Corresponding halo distributions at the FT plane are also shown in Figures 13c and 13d, respectively.

Filtering through the halos indicated as d_x or d_y yields the components d_x or d_y. It is thus seen that both the components, either with the same sensitivity (apertures are placed with equal distances along the x and y directions) or with different sensitivities (aperture separations are different) can be obtained from a single double-exposure specklegram.

The Leendertz and Duffy methods can be combined into one—the object is illuminated by symmetric collimated beams, and the specklegram is recorded using an apertured lens. This combination extends the range and sensitivity of in-plane displacement measure-

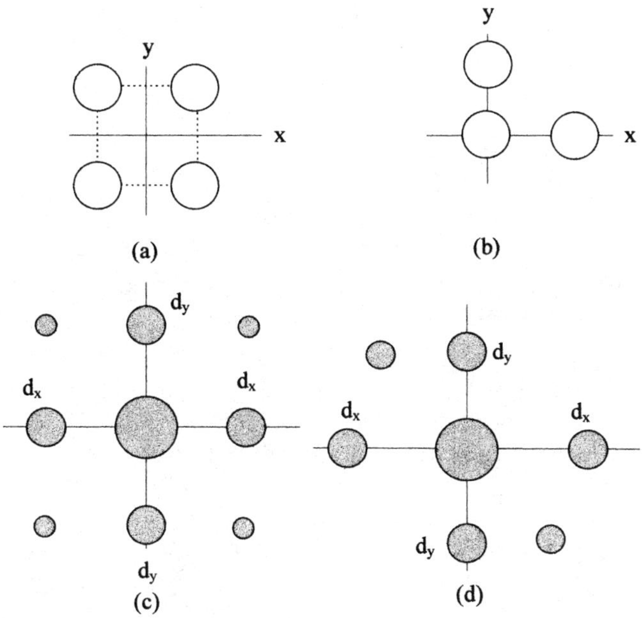

Figure 13 (a,b) Aperture configurations. (c,d) Corresponding halo distributions.

ment from a small value dictated by Leendertz method to a large value governed by Duffy method. When the double-exposure specklegram is filtered, both systems of fringes are observed simultaneously.

XVIII. OUT-OF-PLANE DISPLACEMENT MEASUREMENT

A schematic to measure the out-of-plane component of deformation is shown in Figure 14. The object is illuminated by collimated beam at an angle θ. A two-aperture mask is placed in front of the imaging lens. One of the apertures carries a ground-glass plate that is illuminated by the unexpanded laser beam to generate a diffuse reference wave. The object is imaged via the other aperture. At the image plane, an interference pattern between the diffuse reference

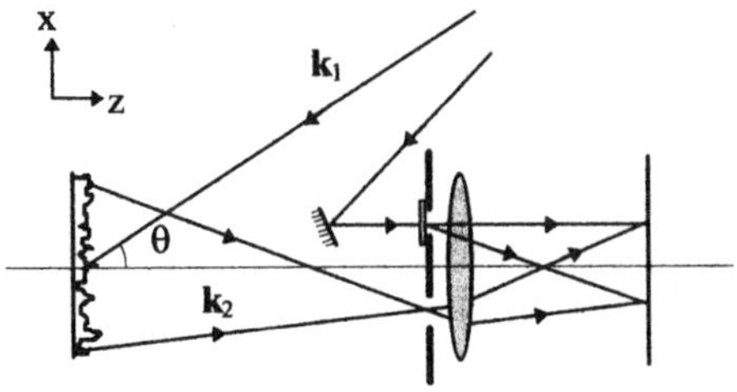

Figure 14 Out-of-plane displacement measurement when $\theta = 0$.

wave and the speckled object image is recorded. This constitutes the first exposure. The object is loaded and the second exposure is made on the same plate. This constitutes a double-exposure specklegram.

On filtering via one of the first-order halos, we obtain a fringe pattern depicting contours of constant out-of-plane displacement. This can be seen as follows: the phase difference δ due to deformation is again given by

$$\delta = (\mathbf{k}_2 - \mathbf{k}_1) \cdot \mathbf{d}$$

where $\mathbf{k}_2 = (2\pi/\lambda)\bar{k}$ and $\mathbf{k}_1 = (2\pi/\lambda)(-\sin\theta\bar{i} + \cos\theta\bar{k})$. Therefore

$$\delta = \frac{2\pi}{\lambda}\{\sin\theta d_x + (1 + \cos\theta)d_z\} \tag{48}$$

The arrangement senses both d_x and d_z components. However, when the object is illuminated normally (i.e., $\theta = 0$), the phase difference δ is $(2\pi/\lambda)2d_z$. The phase difference depends only on the out-of-plane component. The bright fringes are formed wherever

$$\frac{2\pi}{\lambda}2d_z = 2m\pi$$

or

$$d_z = \frac{m\lambda}{2} \tag{49}$$

The consecutive bright fringes are separated by $\lambda/2$ in out-of-plane displacement changes.

XIX. SIMULTANEOUS MEASUREMENT OF OUT-OF-PLANE AND IN-PLANE DISPLACEMENT COMPONENTS

By judicious choice of aperture configuration, it is possible to obtain in-plane and out-of-plane displacement components from a single double-exposure specklegram. One such aperture configuration along with the halo distribution is shown in Figure 15. It may be seen that the halo AB_1 or AB_{-1} yields the y component of the in-plane displacement, while the halo BC_1 or BC_{-1} yields the x

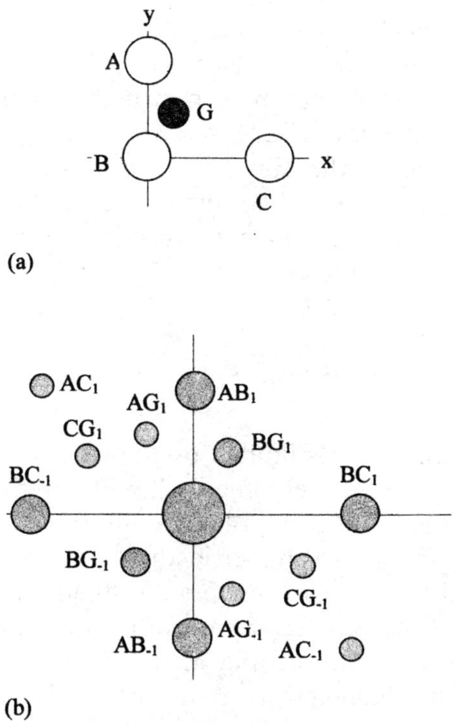

Figure 15 (a) Aperture configuration and (b) halo distribution.

component. The halos BG_1 or BG_{-1} or AG_1 or AG_{-1}, or CG_1 or CG_{-1} yield the z component of the displacement. Obviously, all three components of the displacement vector can be retrieved from a single double-exposure specklegram. In fact, there is redundancy built into the arrangement.

XX. OTHER POSSIBILITIES FOR APERTURING THE LENS

In addition to obtaining all the components of the deformation vector from a single double-exposure specklegram, the aperturing can be used to multiplex the information record. That is, several states of the object can be stored and the information retrieved from the specklegram. The multiplexing can be done by two means, namely frequency modulation (where the apertures are laterally shifted on the lens after each exposure) and θ-multiplexing (modulation), where the apertures are angularly shifted. These methods can also be combined. The amount of information that can be recorded and retrieved depends on the lens size, the aperture size, the spatial frequency content in the object, and the dynamic range of the recording material. In some experiments, in addition to the measurement of displacement components, the slope information is also recorded and retrieved later.

XXI. DUFFY'S ARRANGEMENT: ENHANCED SENSITIVITY

The sensitivity of Duffy's arrangement is limited by the lens aperture. This limitation, however, can be overcome by modifying the recording setup as shown in Figure 16. The object is illuminated normally and is observed along two symmetric directions: the beams are folded and directed to a pair of mirrors and then onto a two-aperture mask in front of the lens. The image of the object is thus made via these folded paths. In this arrangement, the speckle size is governed by the aperture diameter, as is the case with the other methods; the fringe frequency is then determined by the aperture separation, and the sensitivity is governed by the angle

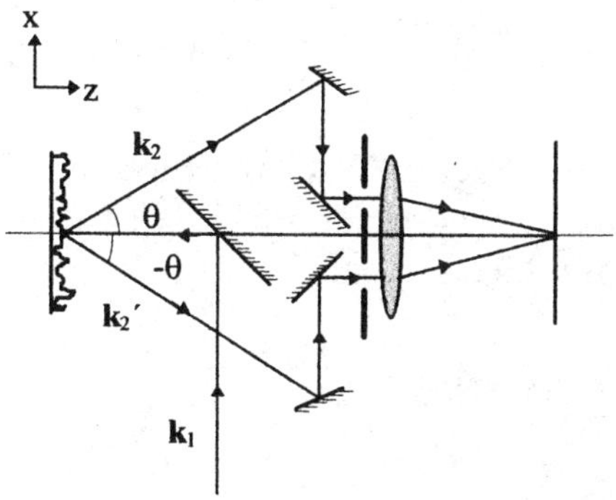

Figure 16 A configuration for in-plane measurement with enhanced sensitivity.

θ, which can be varied over a large range and is not restricted by the lens aperture. The phase difference δ is expressed as follows:

$$\begin{aligned}\delta = \delta_2 - \delta_1 = \delta &= (\mathbf{k}_2 - \mathbf{k}_1)\cdot\mathbf{d} - (\mathbf{k}_2' - \mathbf{k}_1\cdot\mathbf{d} \\ &= (\mathbf{k}_2 - \mathbf{k}_2')\cdot\mathbf{d} = \frac{2\pi}{\lambda}2d_x\sin\theta\end{aligned} \tag{50}$$

The in-plane fringes are extracted by filtering using one of the first-order halos. The technique introduces perspective errors and also the shear when large objects are studied at larger angles. The perspective error, however, can be reduced if a pair of prisms is used to decrease the convergence.

XXII. SPECKLE INTERFEROMETRY: SHAPE MEASUREMENT/CONTOURING

The following methods are available for contouring by means of speckle interferometry:

- change of direction of the illumination beam between exposures
- change of wavelength between exposures: dual-wavelength technique
- change of refractive index of the surrounding medium between exposures: dual-refractive-indices technique

rotation of the object between exposures in an in-plane sensitive configuration.

These methods will be discussed later (Section XXVIII: Electronic Speckle Pattern Interferometry), when we present a technique that allows the contour fringes to be obtained in real time.

XXIII. SPECKLE SHEAR INTERFEROMETRY

So far, we have discussed techniques that measure the displacement components only. As mentioned earlier, a stress analyst is usually interested in strains rather than displacements. Strain is obtained by fitting the displacement data numerically and then differentiating, a procedure that can lead to large errors. Therefore, methods are investigated that can yield fringe patterns that represent the derivatives of the displacement. This is achieved with speckle shear interferometry. Since all the speckle techniques for displacement measurement use subjective speckles (i.e., image plane recordings), we restrict ourselves to shear at the image plane. The shear methods are grouped under the five categories listed in Table 1.

A. The Meaning of Shear

Shear essentially means shift. When an object is imaged via two identical paths and the images are perfectly superposed, there is no shear even though there are two images. Since the imaging is via two independent paths, the images can be independently manipulated. In linear shear, one image is laterally shifted in any desired direction by an appropriate amount. In interferometry, one image acts as a reference to the other; hence there is no need to supply an additional reference wave. When an object is sheared, we compare its response to the external agencies at any point to that of a shifted point.

In rotational shear, one image is rotated, usually about the optical axis, by a small angle with respect to the other image. In radial shear, one image is either radially contracted or expanded with respect the other. The inversion shear allows a point at (x, y) to be compared with another point at $(-x, -y)$. This is equivalent

Table 1 Shear Methods Used in Speckle Interferometry

Shear types	Fringe formation
Lateral shear or linear shear	$\delta(x + \Delta x, y + \Delta y) - \delta(x, y)$
Rotational shear	$\delta(r, \theta + \Delta\theta) - \delta(r, \theta)$
Radial shear	$\delta(r \pm \Delta r, \theta) - \delta(r, \theta)$
Inversion shear	$\delta(x, y) - \delta(-x, -y)$
Folding shear	$\delta(x, y) - \delta(-x, y)$: folding about y axis
	$\delta(x, y) - \delta(x, -y)$: folding about x axis

to a rotational shear of π. In folding shear, a point is compared with its mirror image: the image may be taken about the y axis or the x axis.

XXIV. METHODS OF SHEARING

One of the most commonly used methods of shearing employs the Michelson interferometer, where the object is seen via two independent paths, $OABAD$ and $OACAD$, as shown in Figure 17a. When the mirrors M_1 and M_2 are normal to each other, and at equal distances from A, the two images are perfectly superposed. Tilting one of the mirrors, as shown, displaces one image in the direction of the arrow: the images are linearly separated. A detailed analysis of this arrangement reveals that the use of a beam-splitting cube as the beams travel in glass a distance of three times the size of the cube introduces a large spherical aberration.

All the shear methods described in Table 1 can be conveniently introduced when an aperture mask is placed in front of an imaging lens. For linear shear, a pair of plane parallel plates, wedges (Fig. 17b), or a biprism have been used. Shearing has also been done with gratings. If the imaging lens is cut into two halves, which can be translated in its own plane or along the axis, it makes an excellent shearing device: both the functions of shearing and imaging are performed by the same device. In fact, a diffractive optical element can be designed that performs the functions of both imaging and shearing. Figure 18 shows schematics for the introduction

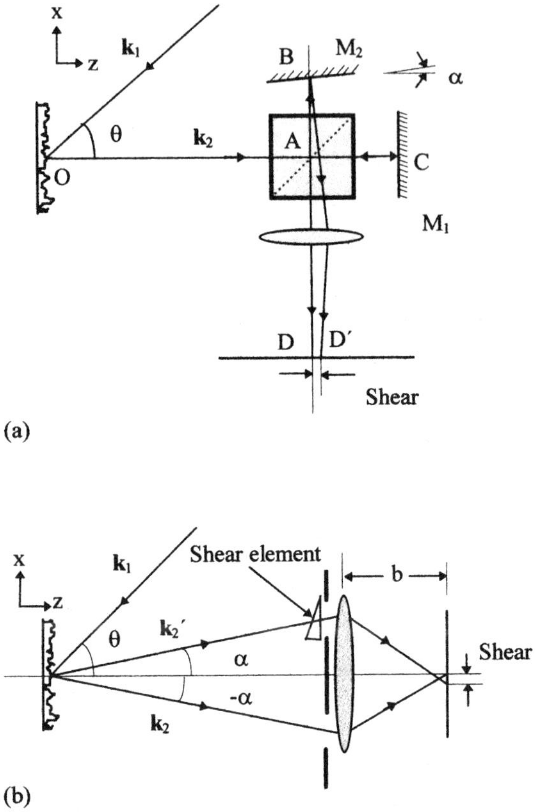

Figure 17 (a) Michelson interferometer for shearing. (b) Duffy's arrangement with a wedge for shearing.

of various shear types by means of an aperture mask and additional optical components.

We now present configurations based on the Michelson interferometer that can be used to introduce both lateral shear and folding shear. Figure 19a shows a conventional arrangement except that the mirrors are replaced by right-angled prisms. The lateral shear is introduced by the translation of one of the prisms. The scheme in Figure 19b is used for folding shear. It uses only one right-angled prism. Depending on the orientation of the prism, folding about the x or the y axis can be achieved.

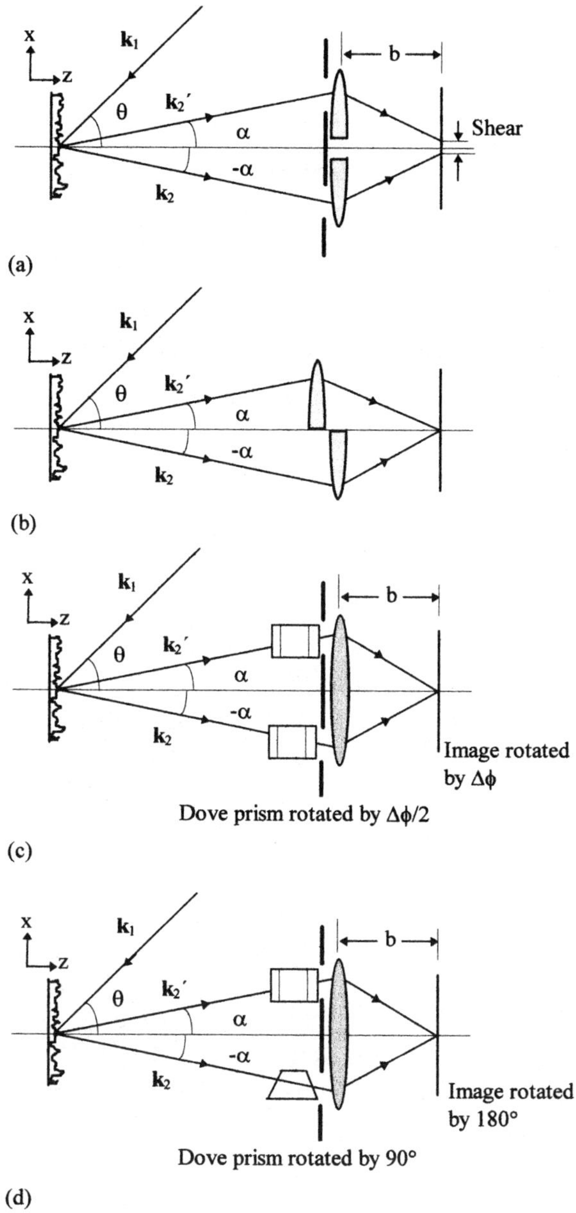

Figure 18 Shearing methods: (a) lateral shear, (b) radial shear, (c) rotational shear, and (d) inversion shear.

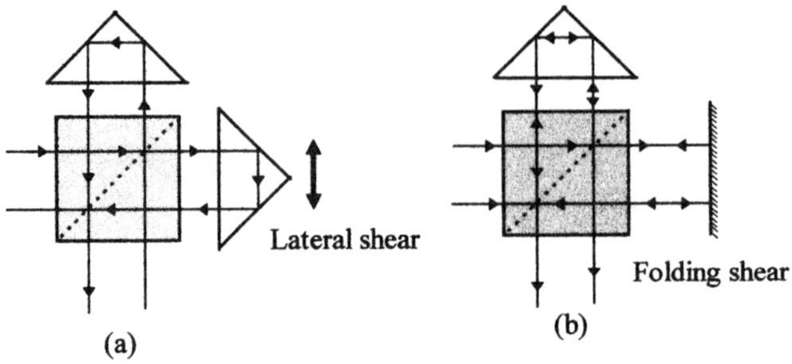

Figure 19 Michelson interferometer for (a) lateral and (b) folding shears.

XXV. THEORY OF SPECKLE SHEAR INTERFEROMETRY

As pointed out earlier, in shear interferometry, a point on the object is imaged as two points or, alternatively, two points on the object are imaged as a single point. One therefore talks of an object plane shear or an image plane shear: these are related through the magnification of the imaging lens.

Let a_1 and a_2 be the amplitudes at any point on the image plane due to two points (x_o, y_o) and $(x_o + \Delta x_o, y_o + \Delta y_o)$ at the object plane. The intensity distribution recorded at the image plane is given by

$$I_1(x, y) = a_1^2 + a_2^2 + 2a_1 a_2 \cos\phi \qquad \text{for } \phi = \phi_2 - \phi_1 \tag{51a}$$

After the object has been loaded, the deformation vectors at the two points are represented by $\mathbf{d}(x_o, y_o)$ and $\mathbf{d}(x_o + \Delta x_o, y_o + \Delta y_o)$, respectively. Therefore, the waves from these two points arrive at the image point with a phase difference $\delta(= \delta_2 - \delta_1)$. The intensity distribution in the second exposure can therefore be expressed as follows:

$$I_2(x, y) = a_1^2 + a_2^2 + 2a_1 a_2 \cos(\phi + \delta) \qquad \text{for } \delta = \delta_2 - \delta_1 \tag{51b}$$

The total intensity recorded is $I_1(x, y) + I_2(x, y)$; hence the amplitude transmittance of the double-exposure specklegram is

$$t(x, y) = t_0 - \beta T\{I_1(x, y) + I_2(x, y)\} \tag{52}$$

On filtering, one obtains a fringe pattern representing the derivatives of the displacement components, as shown later.

XXVI. FRINGE FORMATION

A. The Michelson Interferometer

The phase difference δ can be expressed, assuming shear only along the x direction, as follows:

$$\begin{aligned}\delta &= (\mathbf{k}_2 - \mathbf{k}_1) \cdot \mathbf{d}(x_o + \Delta x_o, y_o) - (\mathbf{k}_2 - \mathbf{k}_1) \cdot \mathbf{d}(x_o, y_o) \\ &\cong (\mathbf{k}_2 - \mathbf{k}_1) \cdot \frac{\partial \mathbf{d}}{\partial x} \Delta x_o\end{aligned} \tag{53}$$

Now, writing for $\mathbf{k}_2$ and $\mathbf{k}_1$ as $\mathbf{k}_2 = (2\pi/\lambda)\bar{k}$ and $\mathbf{k}_1 = (2\pi/\lambda)(-\sin\theta\bar{i} - \cos\theta\bar{k})$ and substituting in Eq. (53) for phase difference, we obtain

$$\delta \cong \frac{2\pi}{\lambda}\left[\sin\theta\frac{\partial d_x}{\partial x} + (1 + \cos\theta)\frac{\partial d_z}{\partial x}\right]\Delta x_o \tag{54}$$

The bright fringes are formed wherever

$$\sin\theta\frac{\partial d_x}{\partial x} + (1 + \cos\theta)\frac{\partial d_z}{\partial x} = \frac{m\lambda}{\Delta x_o} \qquad \text{for } m = 0, \pm 1, \pm 2, \pm 3, \ldots \tag{55}$$

The fringe pattern has contributions from both the strain $(\partial d_x/\partial x)$ and the slope $(\partial d_z/\partial x)$. However, when the object is illuminated normally (i.e., $\theta = 0$), the fringe pattern represents partial x-slope pattern only: that is,

$$\frac{\partial d_z}{\partial x} = \frac{m\lambda}{2\Delta x_o} \tag{56}$$

The fringe pattern corresponding to partial y-slope $(\partial d_z/\partial y)$ is obtained when a shear along the y direction is applied.

B. The Apertured Lens Arrangement

The phase difference δ can again be expressed, assuming shear only along x direction, as follows:

$$
\begin{aligned}
\delta &= (\mathbf{k}_2-\mathbf{k}_1)\cdot\mathbf{d}(x_o+\Delta x_o, y_o) - (\mathbf{k}_2'-\mathbf{k}_1)\cdot\mathbf{d}(x_o, y_o)\\
&\cong (\mathbf{k}_2-\mathbf{k}_1)\cdot\mathbf{d}(x_o, y_o) + (\mathbf{k}_2-\mathbf{k}_1)\cdot\frac{\partial\mathbf{d}}{\partial x}\Delta x_o - (\mathbf{k}_2'-\mathbf{k}_1)\cdot\mathbf{d}(x_o, y_o)\\
&\cong (\mathbf{k}_2-\mathbf{k}_2')\cdot\mathbf{d}(x_o, y_o) + (\mathbf{k}_2-\mathbf{k}_1)\cdot\frac{\partial\mathbf{d}}{\partial x}\Delta x_o\\
\delta &\cong \frac{2\pi}{\lambda}2d_x\sin\alpha + \frac{2\pi}{\lambda}\left[\sin\theta\frac{\partial d_x}{\partial x} + (1+\cos\theta)\frac{\partial d_z}{\partial x}\right]\Delta x_o
\end{aligned}
\tag{57}
$$

Comparison with the expression for the Michelson interferometer reveals that there is an in-plane component-dependent term in addition to the usual expression. This term is due to the two apertures separated by a distance—an arrangement that has been shown to be inherently sensitive to the in-plane component. An interesting aspect of aperturing of a lens, as pointed out earlier, is its ability to measure simultaneously in-plane, out-of-plane displacement components and their derivatives. Figure 20 shows photographs of out-of-plane displacement, partial x-slope and partial y-slope fringe patterns of a defective pipe obtained from the same double-exposure specklegram.

XXVII. SHEAR INTERFEROMETRY WITHOUT THE INFLUENCE OF IN-PLANE COMPONENT

It was shown earlier that shear interferometry performed with an aperture mask in front of the imaging lens always yields a fringe pattern that is due to the combined effect of the in-plane displacement and the derivatives of the displacement. At the same time, it is desirable to have an aperture mask for obtaining high contrast fringe patterns. To retain this advantage and eliminate the in-plane displacement sensitivity, the configuration is modified as shown in Figure 21.

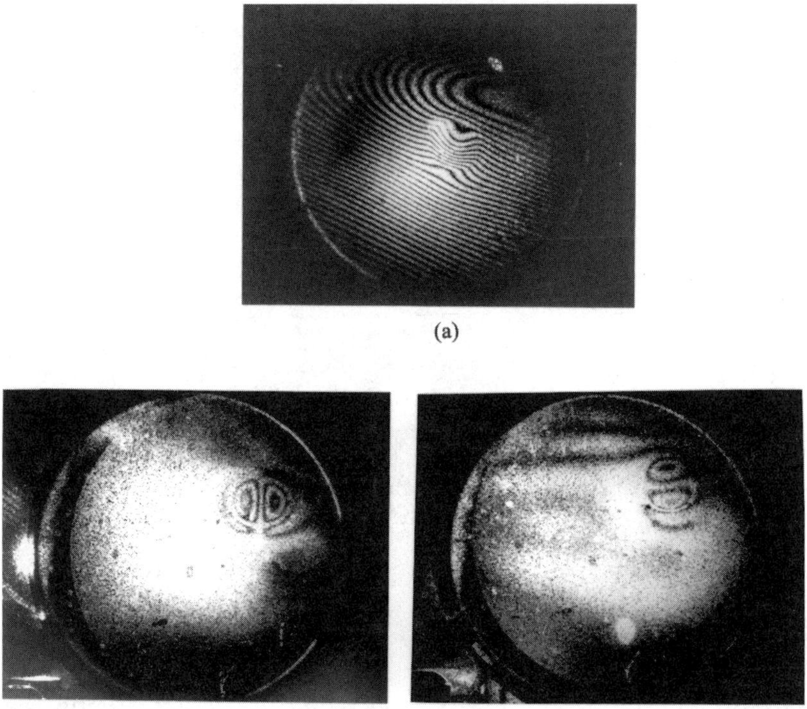

(a)

(b) (c)

Figure 20 (a) Out-of-plane fringe pattern. (b) Partial x-slope fringe pattern. (c) Partial y-slope fringe pattern obtained from the same double-exposure specklegram.

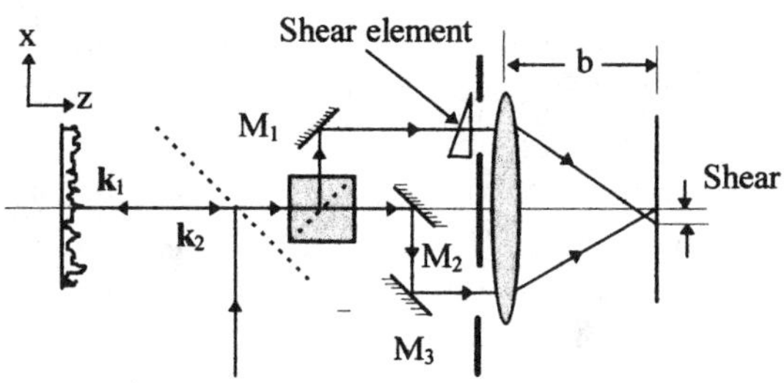

Figure 21 Experimental arrangement insensitive to in-plane displacement in shear interferometry.

The object is illuminated normally and viewed axially. The object beam is divided and arranged to pass through the apertures to form the two images. The shear is introduced by placing a wedge plate in front of an aperture. The shear could just as well be introduced by tilting any one of the mirrors M_1, M_2, or M_3. The mirror combination M_2, M_3 compensates for any extra path difference. The fringe formation is now governed by

$$\frac{\partial d_z}{\partial x} = \frac{m\lambda}{2\Delta x_o} \tag{58}$$

This arrangement gives a pure partial slope fringe pattern.

XXVIII. ELECTRONIC SPECKLE PATTERN INTERFEROMETRY (ESPI)

The speckle size in speckle interferometry can be controlled by the F# of the imaging lens. Further, the size can be doubled by adding a reference beam axially. Thus for recording it is possible to use electronic detectors, which have limited resolution, instead of photographic emulsions. Use of electronic detectors avoids the messy development process, and the processing is done at video rates, making the technique almost possible in real time. Photoemulsions, as mentioned earlier, integrate the light intensity falling on them. In speckle techniques with photographic recording, the two exposures were added in succession, and then techniques were developed to remove the undesired dc component. In electronic detection, the two exposures are independently handled and the subtraction removes the dc component. Phase-shifting techniques are easily incorporated; hence deformation maps can be presented almost in real time. The availability of fast PCs and large-density CCD detectors makes the technique of electronic detection very attractive. In fact, ESPI is an alternative to HI and perhaps will replace it in industrial environments.

One may argue that all the techniques discussed under speckle interferometry and speckle shear interferometry can be adopted simply by replacing the recording medium by the electronic detector. This is indeed not true: since the resolution of the electronic detectors is limited to the range of 50–100 l/mm, speckle sizes in the range of 10–20 μm are desired.

A. Out-of-Plane Displacement Measurement

Figure 22 shows one of the several configurations used for measuring the out-of-plane component. The reference beam is added axially such that it appears to emerge from the center of the exit pupil of the imaging lens. The speckle size is matched with the pixel size by the lens aperture. The intensities of both the object and the reference beams at the CCD plane are adjusted to be equal. The first frame is stored in the frame grabber, and the second frame, captured after the object has been loaded, is subtracted pixel by pixel. The difference signal is rectified and then sent to the monitor to display the fringe pattern. This procedure can be mathematically followed thus:

The intensity recorded in the first frame is given by

$$I_1(x, y) = a_1^2(x, y) + a_2^2(x, y) + 2a_1(x, y)a_2(x, y)\cos\phi \quad \text{for } \phi = \phi_2 - \phi_1 \tag{59a}$$

where we have taken a speckled reference wave. The output of the detector is assumed to be proportional to the intensity incident on it. The intensity recorded in the second frame is

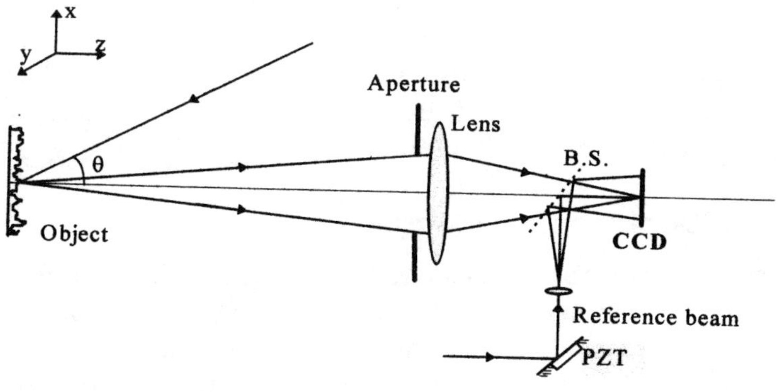

Figure 22 A configuration for ESPI.

$$I_2(x, y) = a_1^2(x, y) + a_2^2(x, y) + 2a_1(x, y)a_2(x, y)\cos(\phi + \delta) \quad (59b)$$

where the phase difference δ introduced by deformation is given by $\delta = (\mathbf{k}_2 - \mathbf{k}_1) \cdot \mathbf{d}$. The subtracted signal is

$$\begin{aligned} \Delta V \propto I_2 - I_1 &= 2a_1(x, y)a_2(x, y)[\cos(\phi + \delta) - \cos\phi] \\ &= 4a_1(x, y)a_2(x, y)\sin\left(\phi + \frac{\delta}{2}\right)\sin\frac{\delta}{2} \end{aligned} \quad (60)$$

The brightness on the monitor will be proportional to the voltage signal ΔV (difference signal) from the detector, hence

$$B = 4\wp\, a_1(x, y)a_2(x, y)\sin\left(\phi + \frac{\delta}{2}\right)\sin\frac{\delta}{2} \quad (61)$$

where $\wp$ is the constant of proportionality. As δ varies, $\sin(\delta/2)$ will vary between -1 and 1. The negative values of $\sin(\delta/2)$ will appear dark on the monitor, resulting in the loss of signal. This loss is avoided by rectifying the signal before it is sent to the monitor. The brightness B is thus given by

$$B = 4\wp\, a_1(x, y)a_2(x, y)\left|\sin\left(\phi + \frac{\delta}{2}\right)\sin\frac{\delta}{2}\right| \quad (62)$$

Brightness will be zero when $\delta/2 = m\pi$ or $\delta = 2m\pi$; $m = 0, \pm1, \pm2, \ldots$. This means that the correlated speckle regions in the speckle pattern would appear dark as a result of the difference operation. This operation also would cause undesirable terms to be eliminated. Phase shifting is easily incorporated by reflecting the reference wave from a PZT-mounted mirror.

B. In-Plane Displacement Measurement

In-plane displacement can be measured using the arrangement due to Leendertz. The sensitivity can be varied by changing the interbeam angle. Unfortunately, the configurations based on aperturing the lens do not work in ESPI because the fringe pattern generated by the pair of apertures is not resolved.

C. Vibration Analysis

ESPI is an excellent tool for studying vibration modes of an object. It can be used to measure extremely small, moderate, and large vibration amplitudes. The arrangement used is the one suited for out-of-plane displacement measurement. The object is excited acoustically or by directly attaching a PZT that is run through a function generator, thereby scanning a large frequency range over which the response of the object could be studied. Since the video rates are much slower than the resonance frequencies, the pattern observed on the monitor represents time-averaged fringes. The intensity distribution is given by $J_0^2\{(4\pi/\lambda)A(x,y)\}$, where $A(x,y)$ is the amplitude of vibration. However, when the reference wave is also modulated at the frequency of object excitation, the intensity distribution in the fringe pattern can be expressed as follows:

$$I(x,y) \propto J_0^2\left\{\frac{4\pi}{\lambda}[A^2(x,y) + a_r^2 - 2A(x,y)a_r\cos(\phi - \phi_r)]^{1/2}\right\} \quad (63)$$

where a_r and ϕ_r are the amplitude and phase of the reference mirror. Obviously, when the object and the reference mirror vibrate in phase, the intensity distribution can be expressed as $J_0^2\{(4\pi/\lambda)[A(x,y) - a)_r]\}$. The zeroth order fringe now occurs where $A(x,y) = a_r$. Therefore, large amplitudes of vibration can be measured. However, if very small vibration amplitudes are to be measured, the frequency of reference wave modulation is taken slightly different from that of the object vibration but still within the video rate. Because of this, the phase of the reference wave varies with time. The intensity distribution can now be expressed as follows:

$$J_0^2\left\{\frac{4\pi}{\lambda}[A^2(x,y) + a_r^2 - 2A(x,y)a_r\cos(\phi - \phi_r(t))]^{1/2}\right\}$$

Since the phase $\phi_r(t)$ varies with time, the argument of the Bessel function varies between $\{A(x,y) + a_r\}$ and $\{A(x,y) - a_r\}$, and the intensity on the monitor will fluctuate. However, if $A(x,y) = 0$, then the argument of the Bessel function remains constant and there is no fluctuation or flicker. Only at locations where the flicker occurs will the amplitude of vibration be nonzero, thereby allowing very small vibration amplitudes to be detected.

D. Shear ESPI Measurement

Again, only the Michelson interferometer-based shear configurations can be used in ESPI. Other methods of shearing produce fringes in speckle too fine to be resolved by the CCD. As mentioned earlier, the fringe pattern carries information about the derivatives of in-plane and out-of-plane components. This can be seen from the expression

$$\delta \cong \frac{2\pi}{\lambda}\left[\sin\theta\frac{\partial d_x}{\partial x} + (1+\cos\theta)\frac{\partial d_z}{\partial x}\right]\Delta x_o$$

where θ is the angle of illumination beam with the z axis. Obviously, pure partial slope fringes are obtained when $\theta = 0$.

When an in-plane-sensitive configuration is used and shear ESPI performed, it is possible to obtain both strain and partial slope fringe patterns. Assuming illumination by one beam at a time, we can express the phase difference introduced by deformation in a shear ESPI setup as follows:

$$\delta_1 \cong \frac{2\pi}{\lambda}\left[\sin\theta\frac{\partial d_x}{\partial x} + (1+\cos\theta)\frac{\partial d_z}{\partial x}\right]\Delta x_o$$

and

$$\delta_2 \cong \frac{2\pi}{\lambda}\left[-\sin\theta\frac{\partial d_x}{\partial x} + (1+\cos\theta)\frac{\partial d_z}{\partial x}\right]\Delta x_o$$

Obviously, when we subtract these two expressions, we obtain strain fringes; if we add, we obtain partial slope fringes.

XXIX. CONTOURING IN ESPI

All the methods of contouring can be easily incorporated in ESPI. We give a brief account of these methods and show how they can be adopted in ESPI.

A. Change of Direction of Illumination

The object is illuminated by a diverging wave from a point source and a usual ESPI setup is used. The first frame is grabbed and stored. The illumination point source is slightly shifted laterally, to change the direction of illumination. The second frame captured now is subtracted from the stored frame and the contour fringes are displayed on the monitor. The contour interval is given by

$$\frac{\lambda}{2\sin\theta}\frac{L}{\Delta s} = \frac{\lambda}{2\sin\theta\Delta\phi} \tag{64}$$

where Δs is the lateral shift of the point source, which causes an angular shift of $\Delta\phi$, and L is the distance between the point source and object plane.

B. Change of Wavelength

The method of dual wavelengths requires two sources or a single source that can be tuned. Two frames are grabbed and subtracted, each frame with one of the wavelengths of light. The true depth contours Δz are generated by this method, whose separation is

$$\Delta z = \frac{\lambda_{eff}}{2} = \frac{\lambda_1\lambda_2}{2|\lambda_1 - \lambda_2|} \tag{65}$$

C. Change of Medium Surrounding the Object

Sometimes the medium surrounding the object is changed between exposures. Subtraction yields true depth contours. In fact, the method is equivalent to the dual wavelength method as the wavelength in the medium changes. We can thus arrive at the same result by writing the wavelength in terms of the refractive index and vacuum wavelength. The contour interval Δz is given by

$$\Delta z = \frac{\lambda}{2|n_1 - n_2|} = \frac{\lambda}{2\Delta n} \tag{66}$$

where Δn is the change in the refractive index when one medium is replaced by the other.

D. Tilt of the Object

In a new contouring method, applicable to speckle interferometry only, one uses an in-plane-sensitive configuration (i.e., the Leendertz configuration). The object is rotated by a small amount between exposures. This converts the depth information, due to rotation, into an in-plane displacement to which the setup is sensitive. The depth contour Δz is given by

$$\Delta z = \frac{\lambda}{2 \sin \theta \sin \Delta\phi} \cong \frac{\lambda}{2 \sin \theta \Delta\phi} \tag{67}$$

where 2θ is the interbeam angle of the illumination waves and $\Delta\phi$ is the angle of rotation. Several modifications of this technique have been published.

XXX. SPECIAL TECHNIQUES: USE OF RETROREFLECTIVE PAINT

Although speckle techniques can be performed with surfaces without any treatment, treatment of the surfaces does help in several cases. For example, coating a dark surface with a white paint improves the contrast of fringes. However, when a retroreflective paint is used, we can essentially enhance the sensitivity of in-plane measurement by a factor of 2. The sensitivity remains practically unchanged for out-of-plane displacement measurements. Figure 23 is a schematic of the experimental arrangement for measuring an in-plane displacement component. It can be easily shown that the phase change introduced as a result of deformation can be written as follows:

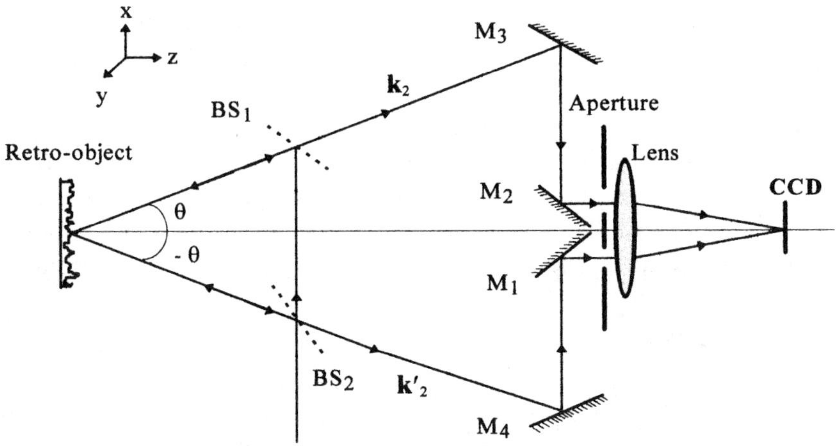

Figure 23 In-plane displacement measurement with enhanced sensitivity in ESPI.

$$\delta = \frac{2\pi}{\lambda} 4d_x \sin\theta$$

This shows that the sensitivity is enhanced by a factor of 2. It may also be seen that one can also use this arrangement for shear ESPI. The shear is introduced by tilt of one of the mirrors. The phase difference introduced is expressed as follows:

$$\delta_2 \cong \frac{2\pi}{\lambda}\left[4d_x \sin\theta + 2\left(\sin\theta\frac{\partial d_x}{\partial x} + \cos\theta\frac{\partial d_z}{\partial x}\right)\Delta x_o\right] \qquad (68)$$

It is seen that it has increased sensitivity for the in-plane derivative, but there is almost no change of sensitivity for the out-of-plane displacement derivative.

XXXI. SPATIAL PHASE SHIFTING

It has been pointed out that if the object wave changes during the time duration required for temporal phase shifting, the results of such a measurement are likely to be grossly erroneous. In such a situation, one uses spatial phase shifting where the information is extracted from a single interferogram. This requires the presence of carrier fringes. Fortunately, the carrier fringes can be easily introduced in the setup due to Duffy. The fringe period can also be varied. Moreover, it has been shown that the sensitivity of Duffy's arrangement can also be increased. Therefore, this arrangement is best suited for spatial phase shifting. It has been used to carry out spatial phase shifting for measuring in-plane and out-of-plane displacement, for shear measurement, and for contouring.

BIBLIOGRAPHY

1. J. C. Dainty (Ed.), *Laser Speckle and Related Phenomena*, Springer-Verlag, Berlin (1975).
2. R. K. Erf (Ed.), *Speckle Metrology*, Academic Press, New York (1978).
3. M. Françon (Ed.), *Laser Speckle and Applications in Optics*, Academic Press, New York (1979).
4. R. S. Sirohi (Ed.), *Selected Papers on Speckle Metrology*, Society of Photo-Optical Instrumentation Engineers Milestone Series MS 35, SPIE Optical Engineering Press, Bellingham, WA (1991).
5. R. S. Sirohi (Ed.), *Speckle Metrology*, Marcel Dekker, New York (1993).
6. P. Meinlschmidt, K. D. Hinsch, and R. S. Sirohi (Eds.), *Selected Papers on Electronic Speckle Pattern Interferometry*, Society of Photo-Optical Instrumentation Engineers Milestone Series MS 132, SPIE Optical Engineering Press, Bellingham, WA (1996).

Chapter 7

Photoelasticity

So far, we have described techniques that are based on interference, or on diffraction of light, and even on the geometrical theory of light. The phenomena of interference and diffraction are manifestations of wave nature, hence are not peculiar to light alone. On the other hand, light is a very small portion of a very wide electromagnetic spectrum. The electromagnetic waves are transverse waves; electric and magnetic field vectors are orthogonal and vibrate perpendicular to the direction of propagation in free space or in an isotropic medium. In fact, electric field vector, magnetic field vector, and propagation vector form an orthogonal triplet. The photoeffects, (seeing, or the recording on a photographic emulsion) are attributed to the electric field vector; hence while dealing with light, we are concerned only with electric field vector (**E** vector).

Let us consider a plane wave propagating along the z direction and the electric field vector confined to the y–z-plane. The tip of the electric field vector describes a line in the y–z-plane as the wave propagates. The y–z plane is called the plane of vibration, and the x–y plane is called the plane of polarization. Such a wave is called a plane-polarized wave. The light emitted by an incandescent bulb or a fluorescent tube is not plane-polarized because the waves emitted by the source, though plane-polarized, are randomly oriented. Such a wave is called unpolarized or natural light. We can obtain polarized light from unpolarized light, and we will discuss this later.

I. SUPERPOSITION OF TWO PLANE-POLARIZED WAVES

Let us consider two orthogonally polarized plane waves propagating in the z direction: in one wave, the **E** vector is vibrating along the y direction and in the other, along the x direction. These waves are described as follows:

$$E_y(z\,;t) = E_{0y}\cos(\omega t - kz + \delta_y) \tag{1}$$

$$E_x(z\,;t) = E_{0x}\cos(\omega t - kz + \delta_x) \tag{2}$$

where δ_y and δ_x are the phases of the waves. These waves satisfy the wave equation. The sum of these waves, due to superposition principle, should also satisfy the wave equation. In general, a wave will have both the x and y components and can be written as follows:

$$\mathbf{E}(z\,;t) = \hat{i}\,E_x(z\,;t) + \hat{j}\,E_y(z\,;t) \tag{3}$$

We wish to find out what track the tip of the electric vector traces when the waves are superposed. To do this, we introduce a variable $\tau = (\omega t - kz)$ and express the plane waves as follows:

$$E_y(z;t) = E_{0y}(\cos\tau\cos\delta_y - \sin\tau\sin\delta_y)$$
$$E_x(z;t) = E_{0x}(\cos\tau\cos\delta_x - \sin\tau\sin\delta_x)$$

From these equations, we obtain

$$\frac{E_y}{E_{0y}}\sin\delta_x - \frac{E_x}{E_{0x}}\sin\delta_y = \cos\tau\,\sin(\delta_x - \delta_y) \tag{4a}$$

$$\frac{E_y}{E_{0y}}\cos\delta_x - \frac{E_x}{E_{0x}}\cos\delta_y = \sin\tau\,\sin(\delta_x - \delta_y) \tag{4b}$$

Squaring both the left-hand and the right-hand sides of Eqs. (4a) and (4b) and summing them, we obtain

$$\left(\frac{E_y}{E_{0y}}\right)^2 + \left(\frac{E_x}{E_{0x}}\right)^2 - 2\frac{E_y}{E_{0y}}\frac{E_x}{E_{0x}}\cos\delta = \sin^2\delta \tag{5}$$

where $\delta = \delta_x - \delta_y$. This equation represents an ellipse. The ellipse is inscribed into a rectangle of sides $2E_{0y}$ and $2E_{0x}$, which are parallel to the y and the x axes. Hence, in the general case of the propagation of a monochromatic wave, the tip of its electric vector traces out an ellipse in any z plane. Such a wave is called elliptically polarized. Since it is a propagating wave, the tip of the E vector traces out a spiral. The tip of the E vector can rotate either clockwise or counterclockwise in the plane. These are termed right-handed polarization (the E vector rotates clockwise when facing the source) and left-handed polarization (the E vector rotates counterclockwise when facing the source of light).

It is shown that the rotation direction is governed by the sign of the phase difference δ. Let us consider a moment of time t_0 when $(\omega t_0 - kz + \delta_y) = 0$. At this moment, $E_y = E_{0y}$ and $E_x = E_{x0}\cos(\delta_x - \delta_y) = E_{x0}\cos\delta$ and $dE_x/dt = -\omega E_{x0}\sin\delta$. The rate of change of E_x (i.e., dE_x/dt) is negative when $0 < \delta < \pi$ and positive when $\pi < \delta < 2\pi$. Obviously, the former case corresponds to the right-polarized wave and the latter to the left-polarized wave.

Two special cases of the elliptical polarization are known as plane polarization, or linear polarization, and circular polarization. Figure 1 shows various polarization states of the light.

II. LINEAR POLARIZATION

When the phase difference between the two waves is a multiple of π, we obtain a linearly polarized wave. When both the waves are in

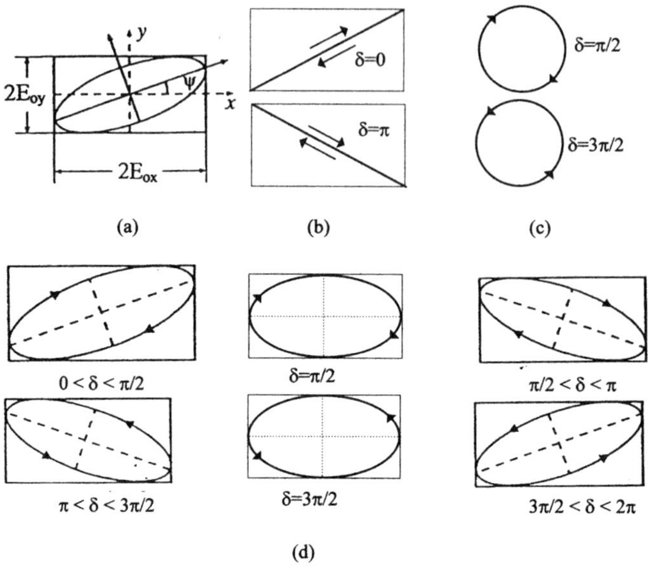

Figure 1 Polarization states of a sinusoidal electromagnetic wave: (a) elliptical, (b) linear, (c) circular, and (d) elliptical, for different δ.

phase (i.e. $\delta = 2m\pi$ for $m = 0, 1, 2 \cdots$), the E vector traces a line in the first and third quadrants. When they are in antiphase [i.e., $\delta = (2m + 1)\pi$], the E vector traces a line in the second and fourth quadrants.

III. CIRCULAR POLARIZATION

When the phase difference between the two waves is an odd multiple of $\pi/2$ [i.e. $\delta = (2m + 1)\pi/2$], the wave is elliptically polarized, but its major and minor axes are now aligned parallel to the x and the y axes. However, if the amplitudes of the two waves are equal, it becomes a circularly polarized wave. It is right-handed circularly polarized if the phase difference is $\pi/2, 5\pi/2, 9\pi/2, \ldots$, and left-handed circularly polarized when $\delta = 3\pi/2, 7\pi/2, 11\pi/2, \ldots$. Therefore, two conditions are to be met to obtain circularly polarized light: the phase difference must be an odd multiple of $\pi/2$ and the amplitudes of both waves must be equal.

IV. PRODUCTION OF POLARIZED LIGHT

Natural light is unpolarized, with the **E** vector taking all possible orientations randomly. We can, however, resolve the **E** vector orientations into two components, one oscillating in the plane of incidence, or p-polarized, and the other orthogonal to this, or s-polarized. These components at any instant have equal amplitudes. It is also seen that we can, in general, obtain an elliptically polarized wave by introducing a phase difference between the orthogonally polarized components. It is therefore required to obtain a linearly polarized wave. Fortunately, there are a number of ways to obtain linearly polarized wave from natural light. These methods are based on:

reflection at a dielectric interface
refraction at a dielectric interface
double refraction
dichroism
scattering

Of these, the first four methods are used for making practical polarizers—the devices that produce linearly polarized light waves from natural light.

A. Reflection

When a beam of light is incident at the air–dielectric interface at an incidence angle θ_B, called the Brewster angle, the reflected beam is linearly polarized: the E vector oscillates in the plane of incidence. This is also called p-polarized beam. The transmitted beam is partially polarized. The Brewster angle θ_B is governed by the refractive index n of the dielectric medium, for reflection at air–dielectric interface, by the relation

$$\tan \theta_B = n$$

It may be noted that the reflection at Brewster angle at the air–dielectric interface fixes the direction of vibration of the electric vector in the reflected beam, hence is used for calibration of polarizers, and so on.

B. Refraction

As mentioned earlier, the transmitted beam is partially polarized even when the incidence is at the Brewster angle. The beam has less p component because it has been removed by reflection at the Brewster angle. However, if several successive reflections are allowed on a number of plane parallel plates aligned at the Brewster angle, most of the p-polarized light will be removed by reflection and the transmitted beam is then s-polarized. Such a polarizer, known as a pile-of-plates polarizer, is often used with high power lasers.

C. Double Refraction

There is a class of crystals in which an incident beam decomposes into two linearly orthogonally polarized beams inside the crystal. The structure of the crystal supports two orthogonally polarized beams. These are the anisotropic crystals. In these crystals, there is a direction along which there is no decomposition. This is known as the optic axis. Some crystals have only one optic axis and are called uniaxial crystals, while others have two optic axes and are called biaxial crystals. Since the uniaxial crystals are important from the point of view of polarizers or other polarization components, we discuss them furhter. Two well-known examples of uniaxial crystals are calcite and quartz.

Let us consider a plate of a uniaxial crystal on which a beam of light is obliquely incident. This beam decomposes into two orthogonally polarized beams inside the plate. One beam obeys the Snell's law of refraction and is called the ordinary beam or o-beam, while the other beam does not obey this law and is called the extraordinary or e-beam. The refractive index n_o of the o-beam is independent of the direction of propagation, while that of e-beam varies with the direction, taking extremum values in a direction orthogonal to the optic axis. If $n_o > n_e$, the crystal is negative uniaxial: calcite is one such crystal ($n_o = 1.658$ and $n_e = 1.486$ for the yellow sodium wavelength). The quartz crystal is a positive uniaxial crystal, since $n_e > n_o$ ($n_o = 1.544$ and $n_e = 1.553$). The optic axis is a slow axis in calcite, while the axis orthogonal to this is a fast axis. The **E** vector in the o-beam oscillates in a plane that is perpendicular to the principal section of the crystal. The principal section contains the optic axis and the direction of propagation. The **E** vector of the e-beam lies in the principal plane.

Since there are two linearly polarized beams inside the crystal, it is easy to obtain a linearly polarized beam by eliminating one of these beams. Fortunately, the angle-dependent refractive index of the e-beam and the availability of media of refractive index intermediate between n_o and n_e make it possible to remove the o-beam by total internal reflection when such a device is made out of calcite crystal. One of the early devices based on this principle is a Nicol prism. Its more versatile companion is a Glan–Thompson prism (Fig. 2). It has the two halves that are cemented by Canada balsam. When an unpolarized beam is incident on the polarizer, the outgoing light beam is linearly polarized. However, if a linearly polarized beam is incident on it, the beam will be completely blocked if the transmission axis of the polarizer is orthogonal to this. The polarizers obtained from anisotropic crystals are generally small but have very high extinction ratios. These devices are usually not used for photoelastic work, which requires large polarizers because photoelastic models are usually moderately large.

1. Phase Plates

Besides obtaining polarizers from anisotropic crystals, we can obtain phase plates, which provide a fixed but wavelength-dependent phase difference between the two components. Let us consider a plane parallel plate of a uniaxial crystal with the optic axis lying in the surface of the plate. A linearly polarized beam is incident on this plate normally. This beam is decomposed into two beams,

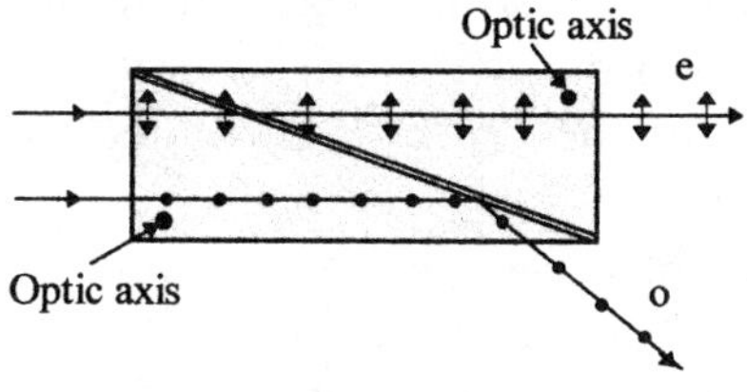

Figure 2 A Glan–Thompson polarizer.

which propagate with different velocities along the slow and fast axes of the plate. Let the refractive indices along these axes be n_o and n_e, respectively. The plate will introduce a path difference $|(n_o - n_e)|\,d$ between the two waves after they have traveled the plate of thickness d. Therefore, any path difference between the two waves can be introduced by an appropriate choice of the plate thickness of a given anisotropic crystal.

2. Quarter-wave Plate

The plate thickness d is chosen to introduce a path difference of $\lambda/4$ or an odd multiple thereof [e.g., $(2m+1)\lambda/4$, where m is an integer]. In other words, such a plate introduces a phase difference of quarter-wave. Therefore

$$d = \frac{(2m+1)}{|n_o - n_e|}\frac{\lambda}{4}$$

For a half-wave plate, the path difference introduced is $\lambda/2$ or $(2m+1)\lambda/2$.

A quarter-wave plate is used to convert the linearly polarized beam into a circularly polarized beam. It is oriented such that the **E** vector of the incident beam makes an angle of 45° with either the fast or the slow axis of the quarter-wave plate. The components in the plate are then of equal amplitudes, and the plate introduces a path difference of $\lambda/4$ between these components. The outgoing beam is thus circularly polarized. The handedness of circular polarization can be changed by rotating the plate by 90° about the axis of the optical beam.

A half-wave plate, on the other hand, rotates the plane of linearly polarized beam. For example, if a beam of linearly polarized light with its azimuth at 45° is incident on a half-wave plate, its azimuth is rotated by 90°. In other words, the beam emerges as linearly polarized but the orientation of the E vector is rotated by 90°.

Phase plates are devices that introduce fixed phase differences. In some applications, we are required either to introduce a path difference that could be varied or to compensate the path difference. This is achieved by compensators. There are two well-known compensators: the Babinet and the Soleil–Babinet compensator. The Babinet compensator consists of two wedge plates with the optic axes orthogonal to each other as shown in Figure 3a. The role of the o- and e-beams changes when the beams enter from one wedge to the other. The path difference introduced by the compensator is given by $|n_o - n_e|(t_2(y) - t_1(y))$, where $t_2(y)$ and $t_1(y)$ are the thicknesses of the two wedge plates. Obviously, the path difference varies along the y direction on the wedge plate. However, if we want a constant path difference between the two beams, we use the Soleil–Babinet compensator. This also consists of two elements—one is a plate, while the other is combination of two identical wedge plates forming a plane parallel plate. The thickness difference is varied by sliding one wedge plate over the other. Thus thickness difference $(t_2 - t_1)$ remains constant over the whole surface. Figure 3b shows the schematic a Soleil–Babinet compensator.

D. Dichroism

Certain anisotropic crystals are characterized by different absorption coefficients with respect to the o- and e-beams. For example, a tourmaline crystal strongly absorbs an ordinary beam. Therefore, we can obtain an e-polarized beam when a beam of natural light passes through a sufficiently thick plate of this crystal. Very large size polarizers based on selective absorption are available as sheets

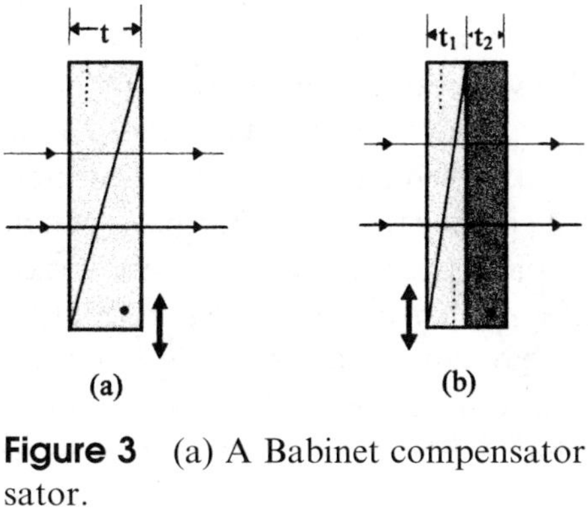

Figure 3 (a) A Babinet compensator and (b) a Soleil–Babinet compensator.

and are known as polaroids or sheet polarizers. These are the ones often used in photoelastic work.

E. Scattering

The light scattered by particles is partially polarized. However, polarizers based on scattering are not in practical use.

V. MALUS' LAW

Consider a linearly polarized light beam incident on a polarizer. The **E** vector of the beam makes an angle θ with the transmission axis of the polarizer. The beam is resolved into two components, one parallel to the transmission axis and the other perpendicular to it. The component perpendicular to the transmission axis is blocked. Therefore, the amplitude of the light transmitted by the polarizer is $E(\theta) = E_o \cos\theta$. Hence, the intensity of the transmitted light is given by $I(\theta) = I_0 \cos^2\theta$, where I_0 is the intensity of the incident beam. This is a statement of Malus' law. It may be seen that a polarizer could also be used as an attenuator in the beam.

VI. THE STRESS–OPTIC LAW

The phenomenon of double refraction or optical anisotropy may also occur in certain isotropic materials, such as glass and plastics, when these are subjected to stress or strain. The condition is temporary and disappears when the stress is removed. This phenomenon was first observed by Brewster and forms the basis of photoelasticity. In photoelasticity, models of objects are cast or fabricated from isotropic materials, which are then subjected to stress. The stresses produce physical deformations, which completely alter the initial isotropic character of the material. We then characterize the material with three principal refractive indices, which lie along the principal axes of the stresses.

The relationship between the principal indices of refraction n_i of a temporary birefringent material and the principal stresses σ_i were formulated by Maxwell and are given by

$$n_1 - n_0 = C_1\sigma_1 + C_2(\sigma_2 + \sigma_3) \tag{6a}$$

$$n_2 - n_0 = C_1\sigma_2 + C_2(\sigma_3 + \sigma_1) \tag{6b}$$

$$n_3 - n_0 = C_1\sigma_3 + C_2(\sigma_1 + \sigma_2) \tag{6c}$$

where n_0 is the refractive index of the unstressed (isotropic) material and C_1, C_2 are constants depending on the material.

For materials under general triaxial stress, the stress–optic law is expressed as follows:

$$n_1 - n_2 = C(\sigma_1 - \sigma_2) \tag{7a}$$

$$n_2 - n_3 = C(\sigma_2 - \sigma_3) \tag{7b}$$

$$n_1 - n_3 = C(\sigma_1 - \sigma_3) \tag{7c}$$

where $C = (C_1 - C_2)$ is the stress–optic coefficient of the photoelastic material.

Let us now consider a plate of an isotropic material of thickness t. This could be subjected to either (a) uniaxial stress or (b) biaxial state of stress. When the plate is subjected to a uniaxial state of stress, $\sigma_2 = \sigma_3 = 0$, hence $n_2 = n_3$. The stress-optic law takes a very simple form:

$$n_1 - n_2 = C\sigma_1 \tag{8}$$

The plate behaves like a uniaxial crystal. When the plate is subjected to a biaxial state of stress (i.e., $\sigma_3 = 0$), the stress–optic law takes the form

$$n_1 - n_2 = C(\sigma_1 - \sigma_2) \tag{9a}$$

$$n_2 - n_3 = C\sigma_2 \tag{9b}$$

$$n_1 - n_3 = C\sigma_1 \tag{9c}$$

The plate behaves like a biaxial crystal.

Now let us consider that a beam of linearly polarized light of wavelength λ is incident normally on a plate of thickness d of the photoelastic material. Within the plate, two linearly polarized beams, one vibrating in the x–z plane and other in y–z plane, are supported. These two waves acquire a phase difference traversing the plate and emerge at the exit surface with a phase difference δ, which is given by

$$\delta = \frac{2\pi}{\lambda}|n_1 - n_2|\,d = \frac{2\pi C}{\lambda}(\sigma_1 - \sigma_2)\,d \tag{10}$$

The phase change δ depends linearly on the difference of the principal stresses and on the thickness of the plate, and inversely on the wavelength of light used. If the beam strikes the plate at an angle i, the phase difference is given by

$$\delta = \frac{2\pi C}{\lambda}\left(\sigma_1 - \sigma_2 \cos^2 i\right) d \sec i \tag{11}$$

In photoelastic practice, it is more convenient to write Eq. (10) in the form:

$$(\sigma_1 - \sigma_2) = \frac{m f_\sigma}{d} \tag{12}$$

where $m(= \delta/2\pi)$ is the fringe order and $f_\sigma(= \lambda/C)$ is the material fringe value for a given wavelength of light. This relationship is known as stress–optic law.

The principal stress difference $(\sigma_1 - \sigma_2)$ in a two-dimensional model can be determined by measuring the fringe order m, if the material fringe value f_σ of the material is known or obtained by calibration. The fringe order at each point in the photoelastic model can be measured by observing the model in a polariscope.

At this juncture, we mention that a plate of thickness d and refractive index n_0 introduces a phase delay of $k(n_0 - 1)d$; $k = 2\pi/\lambda$. When the plate is stressed, the linearly polarized components travel with different speeds and acquire the phase delays $k(n_1 - 1)d_1$ and $k(n_2 - 1)d_1$, where d_1 is the thickness of the stressed plate and is related to the thickness d of the unstressed plate by

$$d_1 = d\left[1 - \frac{\nu}{E}(\sigma_1 + \sigma_2)\right] \tag{13}$$

This change of thickness is very important in interferometry and also in holophotoelasticity.

VII. THE STRAIN–OPTIC LAW

The stress–strain relationships for a material exhibiting perfectly linear elastic behavior under two-dimensional state of stress are

$$\varepsilon_1 = \frac{1}{E}(\sigma_1 - \nu\sigma_2) \tag{14a}$$

$$\varepsilon_2 = \frac{1}{E}(\sigma_2 - \nu\sigma_1) \tag{14b}$$

where E and ν are the Young's modulus and Poisson's ratio of the material, respectively. From Eqs. (14a) and (14b), the principal stress difference is

$$(\sigma_1 - \sigma_2) = \frac{E}{1+\nu}(\varepsilon_1 - \varepsilon_2) \tag{15}$$

Substituting this in the stress–optic law, we obtain

$$(\varepsilon_1 - \varepsilon_2) = \frac{mf_\varepsilon}{d} \tag{16}$$

where $f_\varepsilon\{= [(1+\nu)/E]f_\sigma\}$ is the material fringe value in terms of strain. The relationship given in Eq. (16) is known as the strain-optic law in photoelasticity.

VIII. METHODS OF ANALYSIS

The optical system most often used for stress analysis is the polariscope, a piece of equipment that takes a variety of forms depending on the end use. However, in general, a polariscope consists of a light source, a device to produce polarized light called a polarizer, a model, and a second polarizer called an analyzer. In addition, it may contain a set of lenses, quarter-wave plates, and photographic or recording equipment.

A. Plane Polariscope

The plane polariscope consists of a light source, a light filter, collimating optics to provide a collimated beam, a polarizer, an analyzer, a lens, and photographic equipment, as shown in Figure 4. The model is placed between the polarizer and the analyzer. The polarizer and the analyzer are crossed, thus producing a dark field.

Let the transmission axis of the polarizer be along the y direction. The amplitude of the wave just behind the polarizer is expressed as follows:

$$E_y(z\,;\,t) = E_{0y}\cos(\omega t - kz)$$

where $k = 2\pi/\lambda$. The field incident on the model is also given by this expression, except that z refers to the plane of the model.

Let us also assume that one of the principal stress directions makes an angle α with the transmission direction of the polarizer

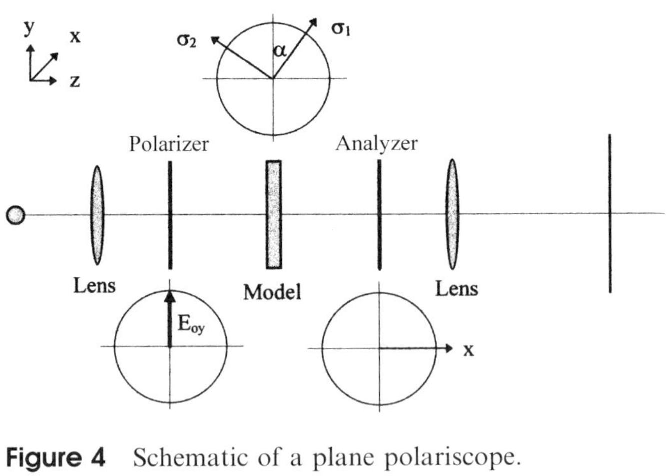

Figure 4 Schematic of a plane polariscope.

(i.e., the y axis). The incident field at the model is split into two components that are orthogonally polarized and vibrate in the planes of σ_1 and σ_2. The amplitudes of these components are $E_{0y} \cos\alpha$ and $E_{0y} \sin\alpha$, respectively. The amplitudes of these components at the exit face of the model are

$$\begin{aligned} E_1(z\,;t) &= E_{0y} \cos\alpha \, \cos[\omega t - kz - k(n_1 - 1)d] \\ &= E_{0y} \cos\alpha \, \cos(\omega t - kz + \delta_y) \end{aligned} \tag{17a}$$

$$\begin{aligned} E_2(z\,;t) &= E_{0y} \sin\alpha \, \cos[\omega t - kz - k(n_2 - 1)d] \\ &= E_{0y} \sin\alpha \, \cos(\omega t - kz + \delta_y - \delta) \end{aligned} \tag{17b}$$

The model introduces a phase difference $\delta = (2\pi/\lambda)(n_1 - n_2)d = (2\pi/f_\sigma)(\sigma_1 - \sigma_2)d$ between these components. The analyzer resolves these components further into components along and perpendicular to its direction of transmission, which is along the x direction. The components along the direction of transmission are allowed through and produce a photoelastic pattern, while those in the orthogonal direction are blocked. The net transmitted amplitude is

$$\begin{aligned} E_1 \sin\alpha - E_2 \cos\alpha &= \frac{E_{0y}}{2} \sin 2\alpha \, [\cos(\omega t - kz + \delta_y) \\ &\quad - \cos(\omega t - kz + \delta_y - \delta)] \\ &= E_{0y} \sin 2\alpha \, \sin\frac{\delta}{2} \, \sin\left(\omega t - kz + \delta_y - \frac{\delta}{2}\right) \end{aligned} \tag{18}$$

This also represents a wave of amplitude $E_{0y} \sin 2\alpha \sin(\delta/2)$ propagating along the z direction. The intensity of this wave therefore is

$$I = E_{0y}^2 \sin^2 2\alpha \, \sin^2\frac{\delta}{2} = I_0 \sin^2 2\alpha \, \sin^2\left[\frac{\pi(\sigma_1 - \sigma_2)\, d}{f_\sigma}\right] \tag{19}$$

The intensity of the transmitted beam is governed by α, the orientation of the principal stress direction with respect to the polarizer transmission axis, and the phase retardation δ. The transmitted intensity will be zero when $\sin 2\alpha \sin(\delta/2) = 0$. In other words, the transmitted intensity is zero when either $\sin 2\alpha = 0$ or $\sin(\delta/2) = 0$. When $\sin 2\alpha = 0$, the angle $\alpha = 0$ or $\alpha = \pi/2$. In either case, one of the principal stress directions is aligned with the polarizer transmission axis. Therefore, these dark fringes give the directions of the principal stresses at any point on the model and are known as isoclinics or isoclinic fringes. When $\sin(\delta/2) = 0$, then $\delta = 2m\pi$, or

$$\sigma_1 - \sigma_2 = \frac{mf_\sigma}{d} \tag{20}$$

The transmitted intensity is zero when $(\sigma_1 - \sigma_2)$ is an integral multiple of f_σ/d. Therefore, the fringes are loci of constant $(\sigma_1 - \sigma_2)$ and are referred to as isochromatics. The adjacent isochromatics differ by f_σ/d. When white light is used for illumination of the model, these fringes are colored; each color corresponds to a constant value of $(\sigma_1 - \sigma_2)$; hence the name isochromatics.

It is seen that isoclinics and isochromatics appear simultaneously in a plane polariscope. It is desirable to separate these fringe patterns. A circular polariscope accomplishes this function and provides only isochromatics.

The isoclinics are the loci of points at which the directions of the principal stresses are parallel to the transmission axes of the polarizer and the analyzer. The isoclinic pattern is independent of the magnitude of the load applied to the model and the material fringe value. When white light is used for illumination, the isoclinics appear dark in contrast to the isochromatics, which except for the zeroth-order fringe are colored. In regions where the directions of the principal stresses do not vary greatly from point to point, the isoclinics appear as wide diffuse bands. The isoclinics do not intersect each other except at an isotropic point, at which the principal stresses are equal in magnitude and sign: that is, $(\sigma_1 - \sigma_2) = 0$. Further, at a point on a shear-free boundary, where the stress parallel to the boundary has a maximum or a minimum value, the isoclinic intersects the boundary orthogonally.

B. Circular Polariscope

It may be seen that isoclinic fringes appear because a linearly polarized light wave is incident on the model. They will disappear if the light incident on the model is circularly polarized. Therefore, a circular polarizer, which is a combination of a linear polarizer and a quarter-wave plate at 45° azimuth, is required. Further, to analyze this light, we also need a circular analyzer. Therefore, a circular polariscope consists of a light source, collimating optics, a polarizer, two quarter-wave plates, an analyzer, and the recording optics, as shown in Figure 5. The model is placed between the two quarter-wave plates. Since these plates are designated with slow and fast axes, they could be arranged in two ways, namely, axes parallel and axes crossed. Similarly, the transmission axes of the polarizer and the analyzer can be parallel or crossed. Therefore, there are four ways of assembling a circular polariscope: two of these possibilities give a dark field at the output, and the remaining two, a bright field. We now consider a configuration in which the analyzer and the polarizer are crossed and the quarter-wave plates are in parallel, resulting in a bright-field output.

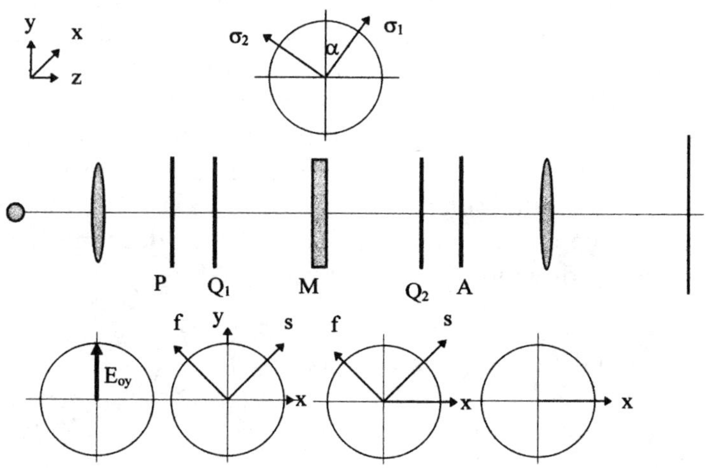

Figure 5 Schematic of a circular polariscope: P, polarizer; Q_1, Q_2, quarter-wave plates; M, model; A, analyzer; for other symbols, see text.

Let the transmission axis of the polarizer be along the y direction. The field transmitted by the polarizer is represented by

$$E_y(z\,;\,t) = E_{0y}\cos(\omega t - kz)$$

This field is split into two components that propagate along the fast and the slow axes of the quarter-wave (QWP) plate. The field at the exit face of first QWP is

$$E_1(z\,;\,t) = \frac{E_{0y}}{\sqrt{2}}\cos\left[\omega t - kz - k(n'-1)\,d'\right] = \frac{E_{0y}}{\sqrt{2}}\cos\left(\omega t - kz + \psi_1\right) \tag{21a}$$

$$\begin{aligned} E_2(z\,;\,t) &= \frac{E_{0y}}{\sqrt{2}}\cos\left[\omega t - kz - k(n''-1)\,d'\right] \\ &= \frac{E_{0y}}{\sqrt{2}}\cos\left(\omega t - kz + \psi_1 - \frac{\pi}{2}\right) = \frac{E_{0y}}{\sqrt{2}}\sin\left(\omega t - kz + \psi_1\right) \end{aligned} \tag{21b}$$

where the phase difference $\pi/2$, introduced by the QWP of thickness d', is expressed as $(2\pi/\lambda)(n'' - n')d' = \pi/2$ and $\psi_1 = -(2\pi/\lambda)(n' - 1)d'$. This signifies that n' corresponds to the fast axis of the QWP. This field is now incident on the model, hence is further decomposed along the directions of the principal stresses. We assume that one of the principal stress directions makes an angle α with the polarizer transmission axis (i.e., the y axis). The field at the entrance face of the model, as decomposed along the σ_1 and σ_2 directions, is now represented by

$$\begin{aligned} E_{\sigma_1}(z\,;\,t) &= \frac{E_{0y}}{\sqrt{2}}\cos\left(\frac{\pi}{4} - \alpha\right)\cos\left(\omega t - kz + \psi_1\right) + \frac{E_{0y}}{\sqrt{2}} \\ &\quad \sin\left(\frac{\pi}{4} - \alpha\right)\sin\left(\omega t - kz + \psi_1\right) \qquad (22a) \\ &= \frac{E_{0y}}{\sqrt{2}}\cos\left(\omega t - kz + \psi_1 - \frac{\pi}{4} + \alpha\right) = \frac{E_{0y}}{\sqrt{2}}\cos\tau \end{aligned}$$

$$E_{\sigma_2}(z\,;t) = -\frac{E_{0y}}{\sqrt{2}}\sin\left(\frac{\pi}{4}-\alpha\right)\cos\left(\omega t - kz + \psi_1\right) + \frac{E_{0y}}{\sqrt{2}}$$
$$\cos\left(\frac{\pi}{4}-\alpha\right)\ \sin\left(\omega t - kz + \psi_1\right) \tag{22b}$$
$$= \frac{E_{0y}}{\sqrt{2}}\sin\left(\omega t - kz + \psi_1 - \frac{\pi}{4} + \alpha\right) = \frac{E_{0y}}{\sqrt{2}}\sin\tau$$

The field amplitudes at the exit face of the model are given by

$$E_{\sigma_1}(z\,;t) = \frac{E_{0y}}{\sqrt{2}}\cos(\tau + \psi_2) \tag{23a}$$

$$E_{\sigma_2}(z\,;t) = \frac{E_{0y}}{\sqrt{2}}\sin(\tau + \psi_2 + \delta) \tag{23b}$$

where $\psi_2 = -(2\pi/\lambda)(n_1 - 1)d$ and $\delta = (2\pi/\lambda)(n_1 - n_2)d$. Now, we decompose these fields along the axes of the second quarter-wave plate, which are inclined at 45° and −45° with the y axis. These are expressed as follows:

$$E_1'(z\,;t) = \frac{E_{0y}}{\sqrt{2}}\cos\left(\frac{\pi}{4}-\alpha\right)\cos(\tau + \psi_2)$$
$$-\frac{E_{0y}}{\sqrt{2}}\sin\left(\frac{\pi}{4}-\alpha\right)\sin(\tau + \psi_2 + \delta) \tag{24a}$$

$$E_2'(z\,;t) = \frac{E_{0y}}{\sqrt{2}}\sin\left(\frac{\pi}{4}-\alpha\right)\cos(\tau + \psi_2) + \frac{E_{0y}}{\sqrt{2}}$$
$$\cos\left(\frac{\pi}{4}-\alpha\right)\sin(\tau + \psi_2 + \delta) \tag{24b}$$

We now assume that both the quarter-wave plates are aligned parallel; that is, their fast and slow axes are parallel. The field at the exit face of the second QWP is

$$E_1'(z\,;t) = \frac{E_{0y}}{\sqrt{2}}\cos\left(\frac{\pi}{4}-\alpha\right)\cos(\tau + \psi_2 + \psi_1)$$
$$-\frac{E_{0y}}{\sqrt{2}}\sin\left(\frac{\pi}{4}-\alpha\right)\sin(\tau + \psi_2 + \delta + \psi_1) \tag{25a}$$

$$\begin{aligned} E_2'(z\,;t) &= \frac{E_{0y}}{\sqrt{2}}\sin\left(\frac{\pi}{4}-\alpha\right)\cos\left(\tau+\psi_2+\psi_1-\frac{\pi}{2}\right) \\ &\quad+\frac{E_{0y}}{\sqrt{2}}\cos\left(\frac{\pi}{4}-\alpha\right)\sin\left(\tau+\psi_2+\delta+\psi_1-\frac{\pi}{2}\right) \\ &= \frac{E_{0y}}{\sqrt{2}}\sin\left(\frac{\pi}{4}-\alpha\right)\sin(\tau+\psi_2+\psi_1) \\ &\quad-\frac{E_{0y}}{\sqrt{2}}\cos\left(\frac{\pi}{4}-\alpha\right)\cos(\tau+\psi_2+\delta+\psi_1) \end{aligned} \tag{25b}$$

Since the analyzer is crossed, it takes the components along the x direction. Therefore, the amplitude of the transmitted wave is given by

$$\begin{aligned} E'(z\,;t) \\ &= \frac{E_{0y}}{2}\begin{bmatrix} \cos\left(\frac{\pi}{4}-\alpha\right)\cos(\tau+\psi_2+\psi_1)-\sin\left(\frac{\pi}{4}-\alpha\right)\sin(\tau+\psi_2+\delta+\psi_1) \\ -\sin\left(\frac{\pi}{4}-\alpha\right)\sin(\tau+\psi_2+\psi_1)+\cos\left(\frac{\pi}{4}-\alpha\right)\cos(\tau+\psi_2+\delta+\psi_1) \end{bmatrix} \end{aligned} \tag{26}$$

Equation (26) can be rewritten as follows:

$$\begin{aligned} E'(z\,;t) \\ &= \frac{E_{0y}}{2}\left[\cos\left(\tau+\psi_2+\psi_1+\frac{\pi}{4}-\alpha\right)+\cos\left(\tau+\psi_2+\delta+\psi_1+\frac{\pi}{4}-\alpha\right)\right] \\ &= E_{0y}\cos\left(\frac{\delta}{2}\right)\cos\left(\tau+\psi_2+\frac{\delta}{2}+\psi_1+\frac{\pi}{4}-\alpha\right) \end{aligned} \tag{27}$$

Again, this represents a wave with an amplitude of $E_{0y}\cos(\delta/2)$. Therefore, the transmitted intensity is given by

$$I = I_0\cos^2\frac{\delta}{2} \tag{28}$$

When there is no stress distribution, $\delta = 0$ and the transmitted intensity is maximum and uniform. This therefore represents a bright-field configuration. It can be seen that when the QWPs are crossed, the field exiting from the second QWP is given by

$$\begin{aligned} E_1'(z\,;t) = {} & \frac{E_{0y}}{\sqrt{2}}\cos\left(\frac{\pi}{4}-\alpha\right)\cos\left(\tau+\psi_2+\psi_1-\frac{\pi}{2}\right) \\ & -\frac{E_{0y}}{\sqrt{2}}\sin\left(\frac{\pi}{4}-\alpha\right)\sin\left(\tau+\psi_2+\delta+\psi_1-\frac{\pi}{2}\right) \end{aligned} \tag{29a}$$

$$\begin{aligned} E_2'(z\,;t) = {} & \frac{E_{0y}}{\sqrt{2}}\sin\left(\frac{\pi}{4}-\alpha\right)\cos(\tau+\psi_2+\psi_1) \\ & +\frac{E_{0y}}{\sqrt{2}}\cos\left(\frac{\pi}{4}-\alpha\right)\sin(\tau+\psi_2+\delta+\psi_1) \end{aligned} \tag{29b}$$

The intensity transmitted by the polarizer is now given by

$$I = I_0\sin^2\frac{\delta}{2} \tag{30}$$

This represents the dark field, since the intensity is zero when there is no stress distribution on the model. Equations (28) and (30) show that the intensity of light emerging from the analyzer in a circular polariscope is a function of the difference of principal stresses $(\sigma_1 - \sigma_2)$ only. The isoclinics fringes have been eliminated.

In a dark-field configuration, the dark fringes occur wherever $\delta = 2m\pi$ ($m = 0, 1, 2, \ldots$), and they correspond to the integer isochromatic fringe order $m = 0, 1, 2, 3, \ldots$, respectively. An example of this fringe pattern is shown in Figure 6a. However, for the bright-field configuration, the dark fringes are obtained when $\delta = (2m + 1)\pi$. These corresponds to isochromatic fringes of half-order (i.e. $m = 1/2, 3/2, 5/2, \ldots$). An example of a light-field fringe pattern is shown in Figure 6b.

IX. EVALUATION PROCEDURE

Directions of principal stresses at any point in the model are determined by means of a plane polariscope. The polarizer and the analyzer are rotated about the optical axis until the isoclinic passes

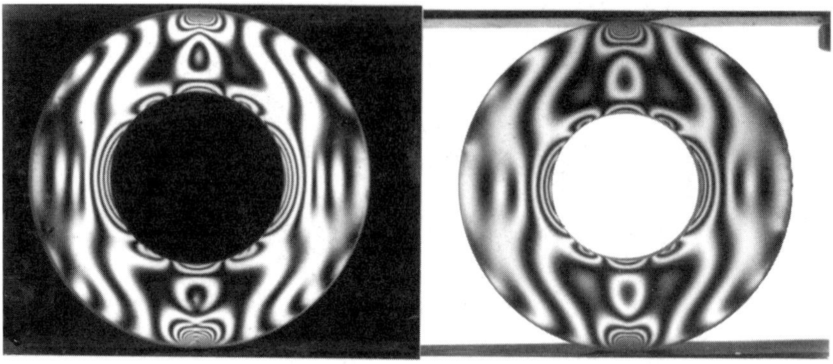

(a) (b)

Figure 6 Isochromatics in (a) dark field, and (b) light field.

through the point of interest. The inclination of the transmission axis gives the principal stress direction. The principal stress difference $(\sigma_1 - \sigma_2)$ in the dark field is given by

$$(\sigma_1 - \sigma_2) = m\frac{f_\sigma}{d} \tag{31a}$$

and in the bright field by

$$(\sigma_1 - \sigma_2) = \left(m + \frac{1}{2}\right)\frac{f_\sigma}{d} \tag{31b}$$

Therefore, the material fringe value f_σ need to be known before $(\sigma_1 - \sigma_2)$ can be calculated. The material fringe order f_σ is obtained by calibration. A circular disc of the same photoelastic material and thickness is used as a model and diametrically loaded. The fringe order in the center of the disc is measured, and f_σ is calculated by means of the formula

$$f_\sigma = \frac{8Fd}{n\pi D} \tag{32}$$

where F is the applied force, D is the diameter of the disc, and n is the measured fringe order at the center of the disc. Other calibration methods use tensile loading or bending.

The principal stress difference $(\sigma_1 - \sigma_2)$ in an arbitrary model is then found by using this value of the material fringe order, and the order n, of the isochromatics. However, the order m is to be counted from $m = 0$, which is normally not known. If it exists in the model, however, it can be easily found by using white light illumination, since $m = 0$ is an achromatic fringe while the higher order fringes are colored. Therefore, a polariscope is usually equipped with both white and monochromatic light sources: the white light source for locating the zeroth-order isochromatic and the monochromatic light source for counting the higher order isochromatics. If the principal stress difference is desired at a point where neither a bright isochromatic nor a dark isochromatic passes, some means of measuring fractional fringe order must be implemented. Methods of measuring fractional fringe order are discussed next.

X. MEASUREMENT OF FRACTIONAL FRINGE ORDER

The methods described here assume that the directions of principal stresses are known. In one method, a Babinet or a Soleil–Babinet compensator is used. The principal axes of the compensator are aligned along the directions of principal stresses. An additional phase difference then can be introduced to shift the dark isochromatics to the point of interest, and this additional phase shift is read from the compensator. Another method, which makes use of a quarter-wave plate for compensation, is known as Tardy's method.

A. Tardy's Method

Tardy's method makes use of a plane polariscope in the dark-field configuration where the intensity distribution is given by $I = I_0 \sin^2 2\alpha \sin^2(\delta/2)$. The observation field contains both isochromatics and isoclinics. Let us now assume that we wish to measure the fractional isochromatics order at a point P as shown in Figure 7. Since it is a dark-field configuration, the integral orders of isochromatics correspond to dark fringes, and to work with the dark fringes we need to brighten the region at and around point P. For this purpose, the polarizer–analyzer combination is rotated by 45° so that their transmission axes make angles of 45° with the directions of the principal stresses. The intensity distribution is now given by $I = I_0 \sin^2(\delta/2)$. A quarter-wave plate is now inserted between the model and the analyzer, with its principal axes parallel to the transmission axes of the polarizer and the analyzer. We can now shift the isochromatics by the rotation of the analyzer by an angle, which is related to the phase shift. To understand the working of this principle, we proceed by expressing the amplitude of the wave transmitted by the polarizer as follows:

$$E_p = E_{op} \cos(\omega t - kz) \tag{33}$$

Since the polarizer–analyzer combination has its transmission axes at 45° with the principal stress directions in the model, this field is resolved along these directions. The components of the field at the exit face of the model are then expressed as follows:

$$E_{\sigma_1} = \frac{E_{op}}{\sqrt{2}} \cos[\omega t - kz - k(n_1 - 1)d] = \frac{E_{op}}{\sqrt{2}} \cos(\omega t - kz + \psi_1) \tag{34a}$$

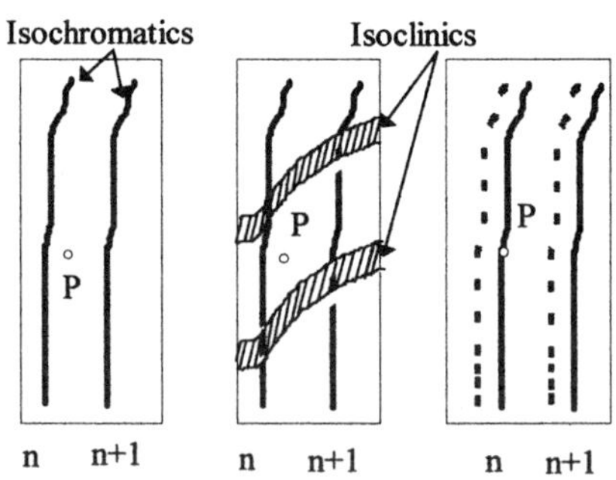

Figure 7 Tardy's method of compensation.

$$E_{\sigma_2} = \frac{E_{\text{op}}}{\sqrt{2}} \cos[\omega t - kz - k(n_2 - 1)d] = \frac{E_{\text{op}}}{\sqrt{2}} \cos(\omega t - kz + \psi_1 + \delta) \tag{34b}$$

where $\delta = k(n_1 - n_2)d$. The field transmitted by the analyzer, which is crossed to the polarizer, is

$$\begin{aligned} E_{\text{A}} &= \frac{E_{\sigma_2}}{\sqrt{2}} - \frac{E_{\sigma_1}}{\sqrt{2}} = \frac{E_{\text{op}}}{2} \left[\cos(\omega t - kz + \psi_1 + \delta) - \cos(\omega t - kz + \psi_1)\right] \\ &= \frac{E_{\text{op}}}{2} 2 \sin\frac{\delta}{2} \sin\left(\omega t - kz + \psi_1 + \frac{\delta}{2}\right) \end{aligned} \tag{35}$$

As mentioned earlier, this describes a wave of amplitude $E_{\text{op}} \sin(\delta/2)$; hence the intensity of the transmitted light is given by $I = I_0 \sin^2(\delta/2)$, as expected. We now introduce a quarter-wave plate after the model and align its axes parallel to the transmission axes of the polarizer and the analyzer. The field components along the fast and slow axes of the quarter-wave plate are

$$E_{\text{f}} = \frac{E_{\sigma_2}}{\sqrt{2}} + \frac{E_{\sigma_1}}{\sqrt{2}} \tag{36a}$$

and

$$E_{\text{s}} = \frac{E_{\sigma_2}}{\sqrt{2}} - \frac{E_{\sigma_1}}{\sqrt{2}} \tag{36b}$$

The field components after passage through the QWP can be expressed as follows:

$$\begin{aligned} E_{\text{f}} &= \frac{E_{\text{op}}}{2} \{\cos[\omega t - kz + \psi_1 - k(n_1' - 1)d'] \\ &\quad + \cos[\omega t - kz + \psi_1 + \delta - k(n_1' - 1)d']\} \\ &= \frac{E_{\text{op}}}{2} [\cos\tau + \cos(\tau + \delta)] \end{aligned} \tag{37a}$$

where $\tau = \omega t - kz + \psi_1 - k(n_1' - 1)d'$ and

$$\begin{aligned} E_s &= \frac{E_{op}}{2}\{\cos[\omega t - kz + \psi_1 + \delta - k(n_2' - 1)d'] \\ &\quad - \cos[\omega t - kz + \psi_1 - k(n_2' - 1)d']\} \\ &= \frac{E_{op}}{2}\left[\cos\left(\tau + \delta - \frac{\pi}{2}\right) - \cos\left(\tau - \frac{\pi}{2}\right)\right] = \frac{E_{op}}{2}[\sin(\tau + \delta) - \sin\tau] \end{aligned} \tag{37b}$$

where $k(n_2' - n_1')d' = \pi/2$.

The analyzer will transmit only E_s component. However, if the analyzer is rotated by an angle χ from this position, the components of the field along the transmission axis of the analyzer are $(E_s \cos\chi + E_f \sin\chi)$. After substitution of E_s and E_f, and a little trigonometry, we obtain the amplitude of the wave transmitted through the analyzer as follows:

$$E_A = E_{op}\sin\left(\chi + \frac{\delta}{2}\right)\cos\left(\tau + \frac{\delta}{2}\right) \tag{38}$$

This represents a wave with an amplitude $E_{op}\sin(\chi + \delta/2)$. Therefore, the intensity of the wave is given by $I = I_0\sin^2(\chi + \delta/2)$. For the mth isochromatic, $\delta = 2m\pi$, hence the intensity at this location must be $I = I_0\sin^2\chi$, which is evidently zero when $\chi = 0$. Further, the intensity at the same location will also be zero when $\chi = \pi$, but then the mth isochromatic would have moved to become the $(m+1)$th isochromatic. In other words, a rotation of the analyzer by π shifts the isochromatics by one order. Therefore, if the analyzer is rotated by an angle χ_p to shift the isochromatics to the point P, the fractional order at that point must be χ_p/π.

XI. PHASE SHIFTING

Phase shifting, a technique for the automatic evaluation of phase maps from the intensity data, was described in detail in Chapter 3. It assumes a special significance in photoelasticity, however, because both beams participating in the interference travel along the same path, and the phase of one cannot be changed independently of other, as is done for the other interferometric methods. We therefore present methods of phase shifting in photoelasticity.

A. Computation of Isoclinics

A bright-field plane polariscope is used to compute isoclinic fringes. The transmitted intensity in this configuration is given by

$$\begin{aligned} I &= I_0 - I_0 \sin^2 2\alpha \ \sin^2\frac{\delta}{2} = I_0\left[1 - \sin^2\frac{\delta}{2}(1 - \cos^2 2\alpha)\right] \\ &= I_0\left[1 - \frac{1}{2}\sin^2\frac{\delta}{2}(1 - \cos 4\alpha)\right] = I_B + V\cos 4\alpha \end{aligned} \tag{39}$$

When the whole polariscope is rotated by β, the intensity transmitted can be expressed as follows:

$$I = I_{\mathrm{B}} + V\cos 4(\alpha - \beta) \tag{40}$$

Both I_{B} and V depend on the value of the isochromatic parameter and consequently on the wavelength of light used. However, the isoclinic parameter does not depend on the wavelength of light. The phase of the isoclinics is obtained by means of a four-step algorithm with the intensity data obtained at $\beta_i = (i - 1)\pi/8$; $i = 1$, 2, 3, and 4, and the relation

$$\tan 4\alpha = \frac{I_4 - I_2}{I_3 - I_1} \tag{41}$$

The phase of the isoclinic fringes is obtained from this relation except at regions where the modulation V is very small. Since V depends on δ, the low modulation areas depend on wavelength. If a monochromatic light is used, there may be several areas in which the value of δ makes the modulation unusable. This problem, however, can be overcome by using a white light source, since the low modulation areas corresponding to a given wavelength will be high modulation areas for another wavelength. Hence, the modulation is kept high enough for use, except at the zeroth-order fringe, where the modulation is obviously zero.

B. Computation of Isochromatics

As can be seen from Tardy's method of compensation, the isochromatic fringe at a point can be shifted by the rotation of the analyzer from the crossed position: a rotation of π shifts the fringe by one order. This presents a nice method of phase shifting, but it suffers from certain drawbacks: the principal axes should be known beforehand, and the isochromatics are calculated for a fixed value of the isoclinic parameter. The phase of the isochromatics can be obtained pixel by pixel by taking intensity data at four positions of the analyzer, say, at 0°, 45°, 90°, and 135°.

We now present a method that is free of the shortcomings just outlined. It takes intensity data at several orientations of a circular polariscope. Table 1 present the configurations, along with the expressions for the transmitted intensity. From these transmitted intensity data, the phase of the isochromatic pattern is computed at each pixel using the relation

Table 1 Expressions for Transmitted Intensity Associated with Eight Polariscope Configurations

No.	Polariscope configuration	Transmitted intensity
1	$P_{\pi/2}Q_{\pi/4}Q_{\pi/4}A_{\pi/4}$	$I_1 = \frac{I_0}{2}(1 + \cos 2\alpha \sin \delta)$
2	$P_{\pi/2}Q_{\pi/4}Q_{-\pi/4}A_{\pi/4}$	$I_2 = \frac{I_0}{2}(1 - \cos 2\alpha \sin \delta)$
3	$P_{\pi/2}Q_{\pi/4}Q_{-\pi/4}A_0$	$I_3 = \frac{I_0}{2}(1 - \cos \delta)$
4	$P_{\pi/2}Q_{\pi/4}Q_{\pi/4}A_0$	$I_4 = \frac{I_0}{2}(1 + \cos \delta)$
5	$P_{-\pi/4}Q_{\pi/2}Q_{\pi/2}A_0$	$I_5 = \frac{I_0}{2}(1 + \sin 2\alpha \sin \delta)$
6	$P_{-\pi/4}Q_{\pi/2}Q_0A_{\pi/2}$	$I_6 = \frac{I_0}{2}(1 - \sin 2\alpha \sin \delta)$
7	$P_{-\pi/4}Q_{\pi/2}Q_0A_{\pi/4}$	$I_7 = \frac{I_0}{2}(1 - \cos \delta)$
8	$P_{-\pi/4}Q_{\pi/2}Q_{\pi/2}A_{\pi/4}$	$I_8 = \frac{I_0}{2}(1 + \cos \delta)$

$$\tan\delta = \frac{(I_1 - I_2)\cos 2\alpha + (I_5 - I_6)\sin 2\alpha}{0.5\{(I_4 - I_3) + (I_8 - I_7)\}} \tag{42}$$

It may be seen that I_3 and I_4 are theoretically equal to I_7 and I_8. Because of polariscope imperfections, however, in practice they may differ; hence all four values are used in the algorithm.

The Fourier transform method can also be used for phase evaluation. The carrier fringe pattern is introduced by a birefringent wedge plate of an appropriate wedge angle. Usually, a carrier frequency of 3–5 l/mm is adequate. The plate is placed close to the model.

XII. BIREFRINGENT COATING METHOD: REFLECTION POLARISCOPE

The use of birefringent coating on the surface of an object extends photoelasticity to the measurement of surface strains on opaque objects and eliminates the need to make models. In this method, a thin layer of a birefringent material is bonded onto the surface of the object. Assuming the adhesion to be good, the displacements on the surface of the object on loading are transferred to the coating, which induces birefringence in the coating. The strain-induced birefringence is observed in reflection. To obtain good reflected intensity, either the surface of the object is polished to make it naturally reflecting or reflective particles are added to the cement that bonds the birefringent coating to the surface of the object. Figure 8 shows a schematic of a reflection polariscope used with birefringent coatings. It can be used as either a plane or circular polariscope. The isochromatics obtained with circular polariscope give the difference between the principal stresses in the coating, that is,

$$(\sigma_1 - \sigma_2)_c = m\frac{f_{\sigma c}}{2d} \tag{43}$$

where d is the thickness and $f_{\sigma c}$ is the fringe value of the coating. Since the light travels through nearly the same region twice, the effective thickness is $2d$. The principal strains are related to the

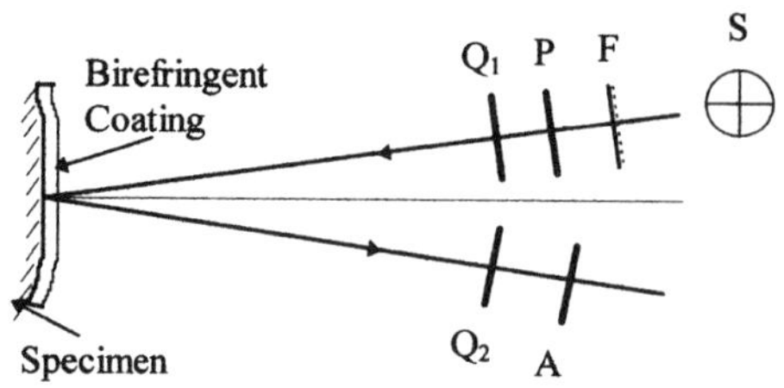

Figure 8 Schematic of a reflection polariscope.

principal stresses through Hooke's law. We thus obtain the difference of the principal strains as

$$\varepsilon_1 - \varepsilon_2 = \frac{1+\nu_c}{E_c}(\sigma_1 - \sigma_2)_c \tag{44}$$

where E_c and ν_c are the elastic constants of the birefringent coating material. Similarly, we could express the difference of principal strains at the surface of the object as follows:

$$\varepsilon_1 - \varepsilon_2 = \frac{1+\nu_o}{E_o}(\sigma_1 - \sigma_2)_o \tag{45}$$

where subscript o stands for object. Assuming that the strains in the coating and at the surface of the object are same, we have

$$(\sigma_1 - \sigma_2)_o = \frac{E_o}{E_c}\frac{1+\nu_c}{1+\nu_o}(\sigma_1 - \sigma_2)_c \tag{46}$$

The separation of stresses in the coating is accomplished by the oblique incidence method. Hence, the principal strains in the coating are calculated. Having obtained the principal strains in the coating, we find the principal stresses at the surface of the object from the following equations:

$$\sigma_{1o} = \frac{E_o}{1-\nu_o^2}(\varepsilon_1 + \nu_o\varepsilon_2) \tag{47a}$$

$$\sigma_{2o} = \frac{E_o}{1-\nu_o^2}(\varepsilon_2 + \nu_o \varepsilon_1) \tag{47b}$$

The analysis is based on the assumption that the strains in the coating and at the surface of the object are the same.

XIII. HOLOPHOTOELASTICITY

The separation of stresses requires that either σ_1, σ_2, or $\sigma_1 + \sigma_2$ be known in addition to $(\sigma_1 - \sigma_2)$ obtained from photoelasticity. The sum of principal stresses is obtained interferometrically, say, by using a Mach–Zehnder interferometer. On the other hand, holophotoelasticity provides fringe patterns belonging to $(\sigma_1 - \sigma_2)$ and $(\sigma_1 + \sigma_2)$ simultaneously, thereby effecting easy separation of stresses. However, the method requires coherent light for illumination. Here, we use holography to record the waves transmitted through the model and later reconstruct this record to extract the information. We can indeed use the technique in two ways. In one method, we obtain only isochromatics (hence the method is equivalent to a circular polariscope); it also provides the flexibility of leisurely evaluation of the fringe pattern. In the other method, both isochromatic and the isopachic fringe patterns are obtained. The method, which requires two exposures, is termed "double-exposure holophotoelasticity", while the former is a single-exposure method.

A. Single-Exposure Holophotoelasticity

The experimental arrangement is shown in Figure 9. The model is already stressed, hence is birefringent. The light from a laser is generally polarized with its **E** vector vibrating in the vertical plane when the beam propagates in the horizontal plane. The beam is expanded and collimated. In case the laser output is randomly polarized, a polarizer is used, followed by a quarter-wave plate oriented at 45°. In brief, the model is illuminated by a circularly polarized wave. The reference wave is also circularly-polarized and of the same handedness, so that both the components are interferometrically recorded.

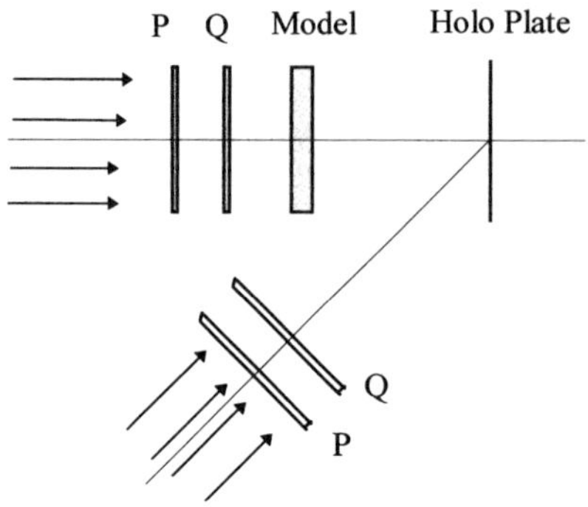

Figure 9 Experimental arrangement for single-exposure holophotoelasticity.

The components of the wave just after the model can be expressed as follows:

$$E_{\sigma_1}(z\,;t) = \frac{E_{0y}}{\sqrt{2}}\cos(\tau + \psi_2) \tag{48a}$$

$$E_{\sigma_2}(z\,;t) = \frac{E_{0y}}{\sqrt{2}}\sin(\tau + \psi_2 + \delta) \tag{48b}$$

Consistent with the treatment presented in Chapter 5 on holography, we write these components as follows:

$$E_{\sigma_1}(z\,;t) = \mathrm{Re}\left\{\frac{E_{0y}}{\sqrt{2}}\exp[i(\tau_1 + \psi_2)]\right\} \tag{49a}$$

$$E_{\sigma_2}(z\,;t) = \mathrm{Re}\left\{\frac{E_{0y}}{\sqrt{2}}\exp\left[i\left(\tau_1 + \psi_2 + \frac{\pi}{2} + \delta\right)\right]\right\} \tag{49b}$$

with $\psi_2 = -(2\pi/\lambda)(n_1 - 1)\,d_1$, $\delta = (2\pi/\lambda)(n_1 - n_2)\,d_1$ and τ_1 does not have the time dependence—which has been ignored because a monochromatic wave is used for illumination—and d_1 is the thickness of the stressed model. Further, $\mathrm{Re}\{\cdot\}$ refers to the real part. Similarly, the reference wave components are written as follows

$$E_{r_1} = \mathrm{Re}\{a_r \exp(i\phi_r)\} \tag{50a}$$

$$E_{r_2} = \mathrm{Re}\left\{a_r \exp\left[i\left(\phi_r + \frac{\pi}{2}\right)\right]\right\} \tag{50b}$$

Since these components are orthogonally polarized, they will interfere with the respective components—essentially, we record two holograms. The recorded intensity is given by

$$I = |E_{\sigma_1} + E_{r_1}|^2 + |E_{\sigma_2} + E_{r_2}|^2 \tag{51}$$

This record, on processing, is a hologram. Assuming linear recording and illumination with the reference beam that releases the two beams, these beams interfere and generate an intensity distribution of the type

$$I = I_0'|\exp[i(\tau_1 + \psi_2)] + \exp[i(\tau_1 + \psi_2 + \delta)]|^2 = I_0(1 + \cos\delta) \tag{52}$$

This is the intensity distribution as obtained in a bright-field circular polariscope. It may be noted that the quarter-wave plate–analyzer combination is not placed behind the model during recording. The state of polarization in the reference wave serves the function of this assembly. If the state of polarization in the reference wave is orthogonal to that in the object wave from the model (i.e., the reference wave is of opposite handedness), the isochromatics pattern corresponding to the dark-field circular polariscope will be obtained.

B. Double-Exposure Holophotoelasticity

The experimental arrangement for double-exposure holophotoelasticity is similar to that shown in Figure 9. The model is illuminated by a circularly polarized wave and another circularly polarized wave of the same handedness is used as a reference wave. The first exposure is made with the model unstressed and the second exposure with the model stressed. During the first exposure, the model is isotropic. However, to be consistent with our earlier treatment, we write the amplitudes of the object and reference waves recorded during the first exposure as follows:

$$E_1(z\,;\,t) = \mathrm{Re}\left\{\frac{E_{0y}}{\sqrt{2}}\exp[i(\tau_1 + \psi_0)]\right\} \tag{53a}$$

$$E_2(z\,;\,t) = \mathrm{Re}\left\{\frac{E_{0y}}{\sqrt{2}}\exp\left[i\left(\tau_1 + \psi_0 + \frac{\pi}{2}\right)\right]\right\} \tag{53b}$$

with $\psi_0 = -(2\pi/\lambda)(n_o - 1)d$ and, again,

$$E_{r_1} = \mathrm{Re}\{a_r \exp(i\phi_r)\} \tag{50a}$$

$$E_{r_2} = \mathrm{Re}\left\{a_r \exp\left[i\left(\phi_r + \frac{\pi}{2}\right)\right]\right\} \tag{50b}$$

In the second exposure, we record two waves from the stressed model. These waves are represented as follows:

$$E_{\sigma_1}(z\,;\,t) = \mathrm{Re}\left\{\frac{E_{0y}}{\sqrt{2}}\exp[i(\tau_1 + \psi_2)]\right\} \tag{54a}$$

$$E_{\sigma_2}(z\,;\,t) = \mathrm{Re}\left\{\frac{E_{0y}}{\sqrt{2}}\exp\left[i\left(\tau_1 + \psi_2 + \frac{\pi}{2} + \delta\right)\right]\right\} \tag{54b}$$

with $\psi_2 = -(2\pi/\lambda)(n_1 - 1)\,d_1$ and $\delta = (2\pi/\lambda)(n_1 - n_2)\,d_1$. The reference waves are the same as those used in the first exposure. As explained earlier, we record two holograms in the second exposure. The total intensity recorded can be written as follows:

$$I = |E_1 + E_{r_1}|^2 + |E_2 + E_{r_2}|^2 + |E_{\sigma_1} + E_{r_1}|^2 + |E_{\sigma_2} + E_{r_2}|^2 \tag{55}$$

The amplitudes of the waves of interest, on reconstruction of the double-exposure hologram, are proportional to

$$2\exp[i(\tau_1 + \psi_0)] + \exp[i(\tau_1 + \psi_2)] + \exp[i(\tau_1 + \psi_2 + \delta)] \tag{56}$$

These three waves interfere to generate a system of isochromatic $(\sigma_1 - \sigma_2)$ and isopachic $(\sigma_1 + \sigma_2)$ fringe patterns. The intensity distribution in the interferogram is given by

$$I = I_0\left\{1 + 2\cos\left(\frac{2\psi_2 + \delta - 2\psi_0}{2}\right)\cos\frac{\delta}{2} + \cos^2\frac{\delta}{2}\right\} \tag{57}$$

Before proceeding further, we need to know what $(2\psi_2 + \delta - 2\psi_0)$ represents. Substituting for ψ_2, δ, and ψ_0 we obtain

$$\begin{aligned} 2\psi_2 + \delta - 2\psi_0 &= -\frac{2\pi}{\lambda}\{2(n_1 - 1)d_1 - (n_1 - n_2)d_1 - 2(n_0 - 1)d\} \\ &= -\frac{2\pi}{\lambda}\{(n_1 + n_2)d_1 - 2n_0 d - 2\Delta d\} \\ &= -\frac{2\pi}{\lambda}\{(n_1 - n_0)d + (n_2 - n_0)d + (n_1 + n_2)\Delta d - 2\Delta d\} \end{aligned} \tag{58}$$

Assuming the birefringence to be small so that $(n_1 + n_2)$ can be replaced by $2n_0$, substituting for $(n_1 - n_0)$ and $(n_2 - n_0)$, and using Eq. (13), we obtain,

$$\begin{aligned} 2\psi_2 + \delta - 2\psi_0 &= -\frac{2\pi}{\lambda}\left\{(C_1 + C_2)(\sigma_1 + \sigma_2)d - 2(n_0 - 1)\frac{\nu}{E}(\sigma_1 + \sigma_2)d\right\} \\ &= -\frac{2\pi}{\lambda}\left\{\left[(C_1 + C_2) - 2(n_0 - 1)\frac{\nu}{E}\right](\sigma_1 + \sigma_2)d\right\} \\ &= -\frac{2\pi}{\lambda}\{(C_1' + C_2')(\sigma_1 + \sigma_2)d\} = \frac{2\pi}{\lambda}C'(\sigma_1 + \sigma_2)d \end{aligned} \tag{59}$$

It is thus seen that the argument of the cosine function in the second term in Eq. (57) depends only on the sum of the principal stresses, hence generates the isopachic fringe pattern. We can rewrite Eq. (57) as follows:

$$\begin{aligned} I = I_0\Big\{&1 + 2\cos\left[\frac{\pi}{\lambda}C'(\sigma_1 + \sigma_2)d\right]\cos\left[\frac{\pi}{\lambda}C(\sigma_1 - \sigma_2)d\right] \\ &+ \cos^2\left[\frac{\pi}{\lambda}C(\sigma_1 - \sigma_2)d\right]\Big\} \end{aligned} \tag{60}$$

It is seen that the second term in Eq. (60) contains information about the isopachics while the second and third terms contain information about the isochromatics. Figure 10 shows an interferogram depicting both types of fringe. We will now examine Eq. (60) and study the formation of isochromatics and isopachics.

Since we are using a bright field configuration, the dark isochromatics would occur when

$$\frac{\pi}{\lambda}C(\sigma_1 - \sigma_2)d = (2n+1)\frac{\pi}{2} \qquad \text{for } n = \text{integer} \tag{61}$$

However, the intensity of the dark isochromatics is not zero but I_0. The bright isochromatics occur when $(\pi/\lambda)C(\sigma_1 - \sigma_2)d = n\pi$, and the intensity in the bright isochromatics is given by

$$I = 2I_0\left\{1 + (-1)^n \cos\left[\frac{\pi}{\lambda}C'(\sigma_1 + \sigma_2)d\right]\right\} \tag{62}$$

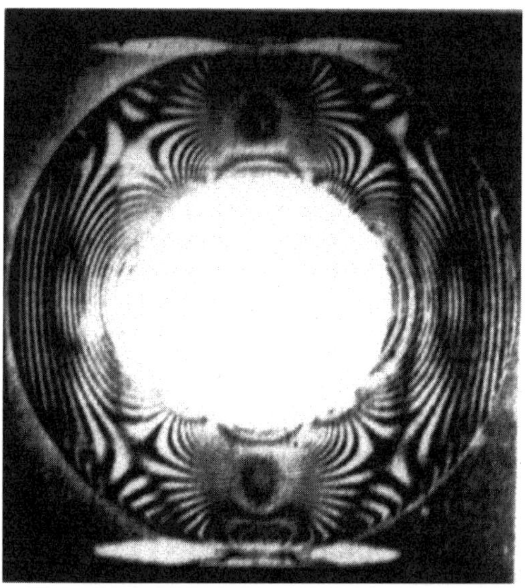

Figure 10 Interferogram showing both the isochromatics (broad fringes) and the isopachics. Notice the phase shift of π when the isopachics cross an isochromatic.

The intensity of the bright isochromatics is modulated by isopachics. Let us first consider bright isochromatics of even order whose intensity is given by

$$I = 2I_0\left\{1 + \cos\left[\frac{\pi}{\lambda}C'(\sigma_1 + \sigma_2)d\right]\right\} \tag{63}$$

The intensity in the bright isochromatics will be zero wherever

$$\frac{\pi}{\lambda}C'(\sigma_1 + \sigma_2)d = (2K + 1)\pi \qquad \text{for } K = 0, 1, 2, 3 \tag{64}$$

The integer K gives the order of isopachics. Wherever this condition is satisfied, the isochromatics would have zero intensity. Therefore, the isopachics modulate the bright isochromatics. Let us now see what happens to bright isochromatics of the next order. Obviously, the intensity distribution in this isochromatics will be

$$I = 2I_0\left\{1 - \cos\left[\frac{\pi}{\lambda}C'(\sigma_1 + \sigma_2)d\right]\right\} \tag{65}$$

If we substitute the condition for the Kth dark isopachics in this equation, we find a maximum intensity of $4I_0$. This shows that the Kth isopachics has changed by one half-order in going over from one bright isochromatic fringe to the next. This interpretation is simple and holds valid when the two families of fringes are nearly perpendicular, as shown in Figure 11. In the other extreme case, where the isochromatics and the isopachics run parallel to each other, the analysis breaks down. It is therefore advisable to separate out these two fringe patterns. The influence of birefringence can be eliminated by passing the beam twice through the model and a Faraday rotator, thereby eliminating the isochromatics pattern. For the model, one can also use materials such as PMMA, which has almost no birefringence, or very little. For such a model, only the isopachic pattern would be observed. Holophotoelasticity can also be performed in real time, which offers certain advantages.

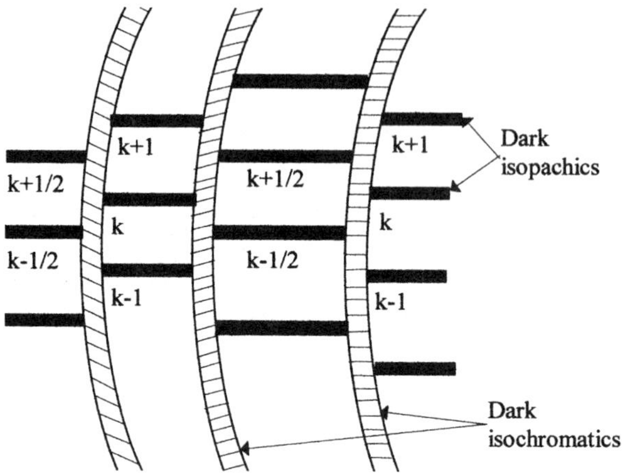

Figure 11 Simplified combined isochromatics and isopachics pattern (drawn).

XIV. THREE-DIMENSIONAL PHOTOELASTICITY

Photoelastic methods, thus far described, cannot be used for investigations of objects under a three-dimensional state of stress. When a polarized wave propagates through such an object, assuming it to be transparent, it integrates the polarization changes over the distance of travel. The integrated optical effect is so complex that it is impossible to analyze it or relate it to the stresses that produce it. There are, however, several methods for investigations of this type. We discuss two such methods, the frozen stress method, which is restricted in its application to static cases of loading by external forces, and the scattered light method.

A. The Frozen Stress Method

The frozen stress method is possibly the most powerful method of experimental stress analysis. It takes advantage of the multiphase nature of the plastics used as model materials. The procedure for stress freezing consists of heating the model to a temperature slightly above the critical temperature and then cooling it slowly to room temperature, typically at less than 2°C per hour under the desired loading condition. The load may be applied to the model either before or after the critical temperature is reached. Extreme care is taken to ensure that the model is subjected to correct loading, since spurious stresses due to bending and gravitational load may be induced because of the low rigidity of the model material at the critical temperature.

After the model has cooled to the room temperature, the elastic deformation responsible for the optical anisotropy is permanently locked. The model is now cut into thin slices for examination under the polariscope. The optical anisotropy is normally not disturbed during slicing if the operation is carried out at high speeds and under coolant conditions. Of the two methods of data collection and interpretation from these slices—subslicing and the oblique incidence method—the latter is the more practical.

B. The Scattered Light Method

When a beam of light passes through a medium containing fine particles dispersed in the volume, a part of the beam is scattered. When the particles are much smaller than the wavelength of light, the intensity of the scattered light varies as ω^4, where ω is the circular frequency of the light waves. This phenomenon was investigated by Rayleigh in detail and is called Rayleigh scattering. The most beautiful observations of the red sunset and blue sky are due to scattering from the gaseous molecules in the atmosphere. Further, the light from the blue sky is partially linear polarized. In some observation directions, the scattered light is linearly polarized.

Consider a scattering center located at a point P as shown in Figure 12. Let the incident light be unpolarized that can be resolved into two orthogonal linearly polarized components with random phases. The incident component vibrating in the y–z plane, when absorbed, will set the particle (rather, the electrons in the particle) vibrating along the y direction. The reradiated wave will have zero amplitude along the y direction. On the other hand, if the particle is oscillating along the x direction, the reradiated wave will have zero amplitude in that direction. Thus, when the observation direction lies along the y direction in the x–y plane

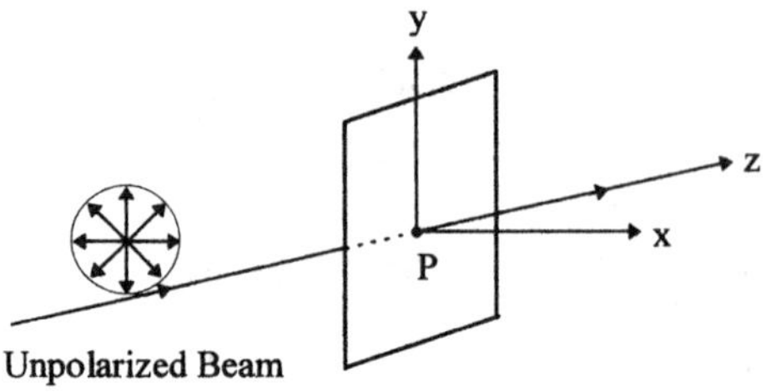

Figure 12 Scattering of unpolarized beam by a scatterer at *P*.

passing through point *P*, the scattered wave will be plane-polarized. The particle acts as a polarizer.

Let us now consider that the incident wave is linearly polarized with the **E** vector vibrating in the *y–z* plane. The electrons in the particle will oscillate along the *y* direction. The reradiated wave will have zero amplitude when observed along the *y* axis. The particle thus acts as an analyzer. This picture is equivalent to placing a polarizer and an analyzer anywhere in the model. Therefore, stress information can be obtained without freezing the stress and slicing the model. The scattered light method therefore provides a nondestructive means of optical slicing in three dimensions.

XV. EXAMINATION OF THE STRESSED MODEL IN SCATTERED LIGHT

A. Incident Light, Unpolarized

Let us consider a stressed model in the path of a narrow beam of unpolarized light. We assume that there are a large number of scatterers in the model. Let us now consider the light scattered by a scatterer at point *P* inside the model when the observation direction is perpendicular to the incident beam. The scattered light is resolved into components along the directions of principal stresses σ_2 and σ_3 as shown in Figure 13. In traversing a distance *PQ* in the model, these two orthogonally polarized components acquire a phase difference. If an analyzer is placed in the observation direction, the transmitted intensity will depend on the phase difference acquired. Since the incident beam is unpolarized, there is no influ-

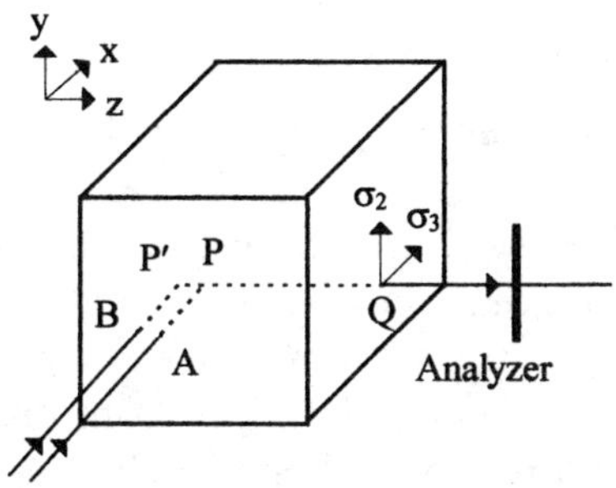

Figure 13 Stressed model: illumination by an unpolarized beam and observation through an analyzer.

ence of traverse AP in the model. Consider that the transmitted intensity is zero for a certain location P of the scatterer; this occurs when the phase difference is a multiple of 2π. As the beam is moved to illuminate another scatterer at point P' in the same plane along the line of sight, the transmitted intensity will undergo cyclic variation between minima and maxima depending on the additional phase acquired when traversing the distance PP'. It is, however, assumed that the directions of principal stresses σ_2 and σ_3 do not change over the distance PP'.

Let m_1 and m_2 be the fringe orders when the light scattered from scatterers at points P and P' is analyzed. Then

$$x_1(\sigma_2 - \sigma_3) = m_1 f \tag{66a}$$

and

$$x_2(\sigma_2 - \sigma_3) = m_2 f \tag{66b}$$

where $x_1 = PQ$ and $x_2 = P'Q$. Therefore, we obtain

$$(\sigma_2 - \sigma_3) = \frac{dm}{dx} f \tag{67}$$

The principal stress difference at any point along the observation direction is proportional to the gradient of the fringe order.

B. Incident Beam, Linearly Polarized

We now consider another situation of a linearly polarized beam incident on the model (see Fig. 14). We assume for the sake of simplicity that the principal stress directions are along the x and the y axes. The transmission axis of the polarizer makes an angle α with the x axis. The incident wave of amplitude E_0 is resolved along the x and the y directions. These linearly polarized components travel with different velocities in the model, hence pick up a phase difference δ. Thus, at any plane normal to the direction of propagation, the state of polarization of the wave in general will be elliptical. It is expressed as follows:

$$\frac{E_x^2}{E_0^2 \cos^2\alpha} + \frac{E_y^2}{E_0^2 \sin^2\alpha} - \frac{2E_x E_y}{E_0^2 \cos\alpha \sin\alpha} \cos\delta = \sin^2\delta \tag{68}$$

where E_x and E_y are the components along the x and the y directions, respectively. The major axis of the ellipse makes an angle of ψ with the x axis where

$$\tan 2\psi = \tan 2\alpha \cos\delta \tag{69}$$

When $\delta = 2p\pi$, $p = 0, \pm 1, \pm 2, \pm 3, \pm 4, \ldots$, the state of polarization of the wave is linear, with orientation $\psi = \pm\alpha$. For positive values of the integer p, the state of polarization of the wave at any plane is the same as that of the incident wave. In general, a scat-

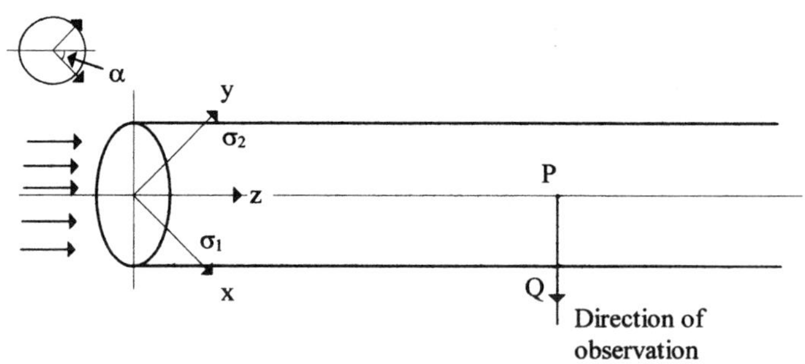

Figure 14 Stressed model: illumination by a linearly polarized beam.

terer at point P in any plane is excited by an elliptically polarized wave. In scattered light photoelasticity, we are looking in the model normal to the direction of propagation; that is, the observation is confined to the plane of the elliptically polarized light. If the observation is made in the direction along the major axis of the ellipse, the amplitude of the reradiated wave received by the observer will be proportional to the magnitude of the minor axis of the ellipse, hence minimum. On the other hand, if the observation direction coincides with the minor axis, the intensity will be maximum. As the beam propagates in the stressed model, the ellipse just described keeps on rotating as the phase difference changes. Therefore, the observation made in the scattered light normal to the direction of the incident beam will show variation in intensity along the length of model in the direction of the incident beam. There is no influence of birefringence in the model on the traverse

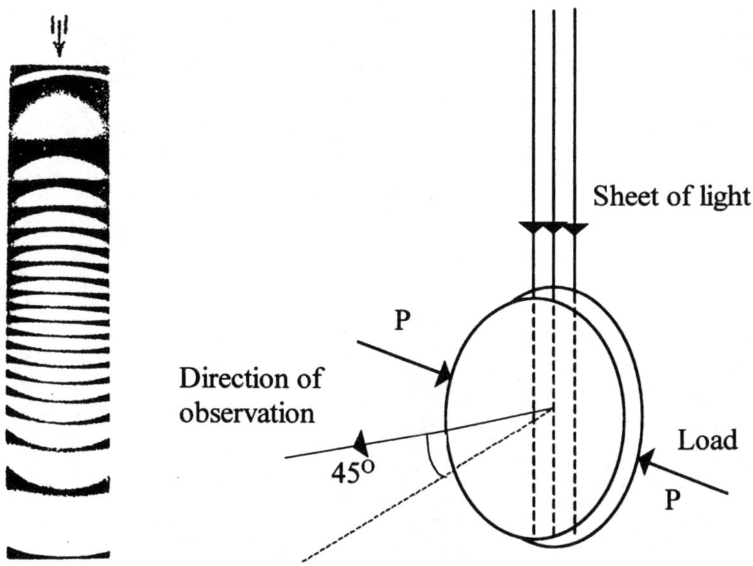

Figure 15 Scattered light pattern of a disc under diametral compression. (b) Schematic showing the directions of illumination and observation.

distance PQ. Figure 15a shows a scattered light stress pattern of a disc under diametral compression. The directions of incident beam and that of the scattered light are shown in Figure 15b.

Owing to the weak intensity of the scattered light, an intense incident beam is used. This beam is realized from a high pressure mercury lamp with suitable collimating optics. The laser is an attractive alternative because the beam can be used without optics. The model is placed in a tank containing an index-matching liquid to avoid refraction and polarization changes at the model surfaces. To facilitate proper adjustments, the model in the tank is mounted on a stage capable of translation and rotation motions.

BIBLIOGRAPHY

1. Albrecht Kuske and George Robertson, *Photoelastic Stress Analysis*, John Wiley, New York (1974).
2. L. S. Srinath, *Scattered Light Photo-elasticity*, Tata McGraw-Hill, New Delhi (1983).
3. Rajpal S. Sirohi and Mahendra P. Kothiyal, *Optical Components, Systems, and Measurement Techniques*, Marcel Dekker, New York (1991).
4. A. S. Kobayashi (Ed.), *Handbook on Experimental Mechanics*, Society for Experimental Mechanics, Bethel, CT (1993).

Chapter 8

The Moiré Phenomenon

When two periodic patterns are superposed, a moiré pattern is formed. It is in fact a very common phenomenon. Moiré patterns are formed by periodic structures of lines or stripes. However, in the general context, the superposed patterns should have opaque and transparent regions. Although moiré patterns can be observed with superposition of periodic objects of nearly the same periodicity, the most commonly used objects are linear gratings with equal opaque and transparent regions, often called Ronchi gratings. Sometimes, circular gratings with equal opaque and transparent regions are used. The moiré phenomenon is a mechanical effect, and the formation of moiré fringes is best explained by mechanical superposition of the gratings. However, when the period in these periodic structures is very fine, diffraction effects play a very significant role.

The fringes formed in holographic interferometry can be considered as a moiré pattern between the primary grating structures belonging to the initial and final states of the object. These primary grating structures are a result of interference between the object wave and the reference wave. Two-wavelength interferometric fringes are also moiré fringes. In general, the moiré pattern can be regarded as a mathematical solution to the interference of two periodic functions. Holo-diagram, a tool developed by Abramson to deal with several issues in HI, is a beautiful device to use in the study of fringe formation and fringe control via the moiré phenomenon.

We can explain the moiré phenomenon by means of either the indicial equation or the superposition of two sinusoidal gratings. We will follow both the approaches, beginning with the indicial equation.

Let us consider a line grating with lines running parallel to the y axis and having a period b (Fig. 1a). This is defined as follows:

$$x = mb \tag{1}$$

where the various lines in the grating are identified by index m, which takes values $0, \pm 1, \pm 2, \pm 3, \ldots$. The second grating of period a is inclined at an angle θ with the y axis as shown in Figure 1b. The lines in this grating are represented by

$$y = x \cot\theta - n\frac{a}{\sin\theta} \tag{2}$$

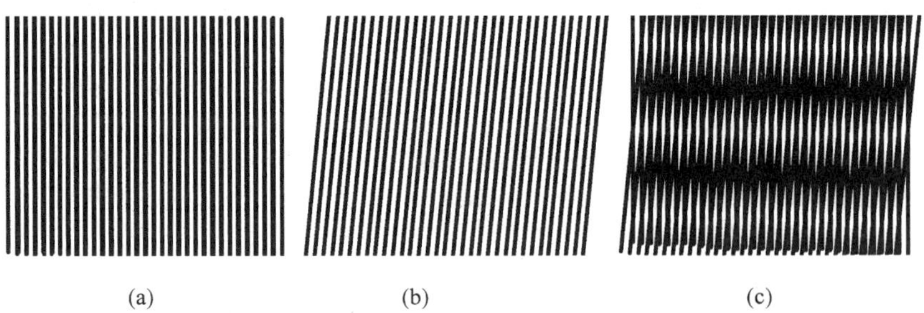

Figure 1 (a) Vertical grating, (b) inclined grating, and (c) moiré pattern as a result of superposition.

where $a/\sin\theta$ is the intercept with the y axis and the index n takes values 0, ±1, ±2, ±3, ..., hence identifies various lines in the grating. On superposing the two line gratings, a moiré fringe pattern is formed and is governed by the indicial equation

$$m \pm n = p \tag{3}$$

where p is another integer taking values 0, ±1, ±2, ±3, ... The plus sign in the indicial equation generates a sum moiré pattern, which usually has high frequency, while the minus sign generates the difference moiré fringes, which are the frequently used and observed moiré patterns. We will use the difference moiré pattern unless otherwise stated. We obtain the equation of the moiré fringes by eliminating m and n from Eqs. (1), (2) and (3) as follows:

$$y = \frac{x(b\cos\theta - a)}{b\sin\theta} + p\frac{a}{\sin\theta} \tag{4}$$

This is shown in Figure 1c. Equation (4) can be written in a more familiar form, like Eq. (2), as follows:

$$y = x\cot\phi + p\frac{d}{\sin\phi} \tag{5}$$

This implies that the moiré pattern is a grating of period d, which is inclined at an angle ϕ with the y axis, where

$$d = \frac{ab}{(a^2 + b^2 - 2ab\cos\theta)^{1/2}} \tag{6a}$$

and

$$\sin\phi = \sin\theta\frac{b}{(a^2 + b^2 - 2ab\cos\theta)^{1/2}} = \frac{d}{a}\sin\theta \tag{6b}$$

It is interesting to obtain moiré patterns for the following two situations,

$a \neq b$ but $\theta = 0$

This is a well-known situation of pitch mismatch: the moiré fringes run parallel to the grating lines. The moiré fringe spacing d is given by $d = ab/|a - b|$. Here $(a - b)$ represents the pitch mismatch. When the gratings are of nearly the same pitch, $a \approx b$, then $d = a^2/|a - b|$. Figures 2a and 2b show gratings of pitches a and b, respectively, with grating elements parallel to the y axis. The moiré fringes due to pitch mismatch are shown in Figure 2c. Physically, the moiré spacing is the distance over which pitch mismatch accumulates to the pitch of the grating itself. When the gratings are of equal period, the moiré spacing is infinite. This arrangement is therefore called as infinite fringe mode.

$a = b$ but $\theta \neq 0$

This case, referred to as an angular mismatch between two identical gratings, results in moiré pattern formation with a period $d = a/2\sin(\theta/2)$, and its orientation ϕ with the y axis is $\pi/2 + \theta/2$.

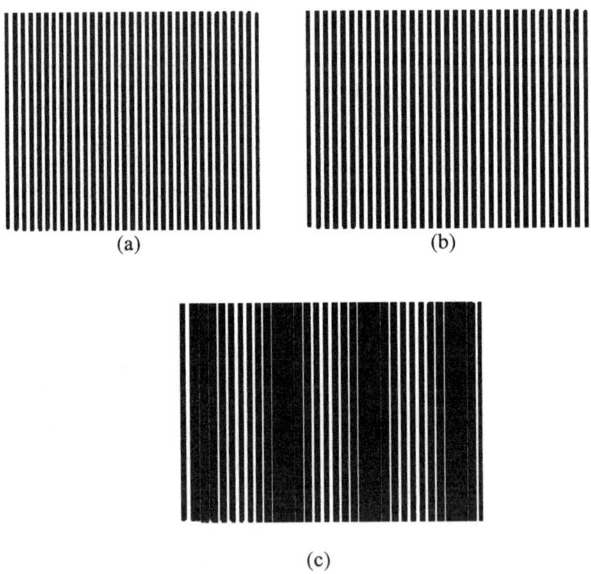

Figure 2 (a) Vertical grating, (b) vertical grating of different pitch, and (c) moiré pattern as a result of pitch mismatch.

In fact, the moiré fringes run parallel to the bisector of the larger enclosed angle between the gratings.

Moiré fringe formation is easily appreciated when we work in the Fourier domain, where every sinusoidal grating can be considered to have three spots lying on a line passing through the origin, with the central spot lying on the origin. This is because the spectrum of a real grating (intensity grating) is centrosymmetric. We can therefore consider only half the spectrum. The spots lie on a line that is perpendicular to the grating elements (lines), and the distance between two consecutive spots is proportional to the frequency of the grating. The second grating also generates its spectrum, which is rotated by an angle equal to the angle between the two gratings. When two gratings are superposed, the difference Δr between the two spots (Fig. 3) generates a grating. If Δr lies within the visibility circle (the spectrum lying in this circle will generate a grating that is seen by an unaided eye), the moiré pattern is formed: the pitch of the moiré fringes is inversely proportional to the length Δr, and the orientation is normal to it. Obviously, when two gratings of equal period are superposed with an angular mismatch, the result is a moiré fringe pattern running parallel to the bisector of these gratings, and the spacing is inversely proportional to the angular mismatch.

I. THE MOIRÉ FRINGE PATTERN BETWEEN A LINEAR GRATING AND A CIRCULAR GRATING

As mentioned earlier, moiré fringes are obtained by the superposition of periodic structures. Since the superposition of a linear grating with a circular grating is of some academic interest, we consider moiré formation by the two gratings. We take a linear grating of

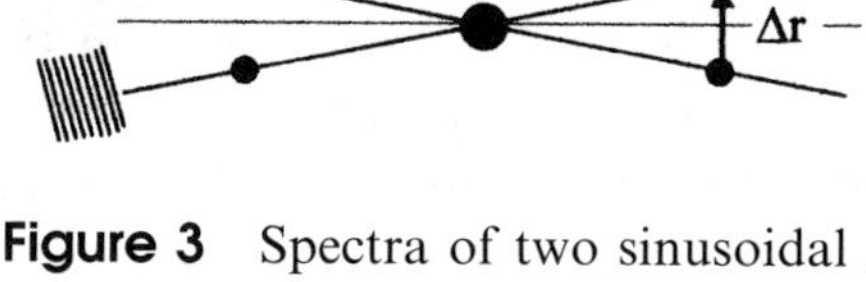

Figure 3 Spectra of two sinusoidal gratings.

period b with its elements running parallel to the y axis. It is represented, again, by

$$x = bm \qquad \text{for } m = 0, \pm 1, \pm 2, \pm 3, \ldots \tag{1}$$

The circular grating of period a is centered at the origin of the coordinate system, hence is represented as follows:

$$x^2 + y^2 = a^2 n^2 \qquad \text{for } n = 0, \pm 1, \pm 2, \pm 3, \ldots \tag{7}$$

The indicial equation is $m \pm n = p$. Therefore, the moiré pattern is given by

$$\frac{x}{b} \pm \frac{(x^2 + y^2)^{1/2}}{a} = p$$

or

$$\frac{x^2 + y^2}{a^2} = p^2 + \frac{x^2}{b^2} - \frac{2xp}{b} \tag{8}$$

This expression can be rewritten as follows:

$$x^2\left(\frac{1}{a^2} - \frac{1}{b^2}\right) + \frac{y^2}{b^2} + 2\frac{x}{b}p - p^2 = 0 \tag{9}$$

Equation (9) represents hyperbolas, ellipses, or parabolas, depending on the relative grating periods. The moiré pattern for $a = b$ appears in Figure 4, which shows parabolic moiré fringes. The Fourier transform of a circular grating lies on a circle; hence an overlap of a linear grating with a circular grating produces a moiré pattern that has a very wide FT spectrum.

II. MOIRÉ BETWEEN SINUSOIDAL GRATINGS

So far, we have used line gratings to look into the formation of moiré patterns. These gratings are seldom used in practice. One normally uses gratings that are binary, since they are very easy to produce. However, when high frequency gratings are used, these

Figure 4 Moiré pattern between a linear grating and a circular grating.

are generally fabricated interferometrically, hence have a sinusoidal profile. Even binary gratings can be Fourier-decomposed into sinusoidal components, however, so it is instructive to see how the moiré of sinusoidal gratings is formed.

The sinusoidal gratings can be recorded either on the same film or on two separate films, which are then overlapped as was done with line gratings. We examine these two situations separately. Let us consider a grating defined by the transmittance function $t_1(x)$:

$$t_1(x) = t_0\left(1 - M\cos\frac{2\pi}{b}x\right) = t_0(1 - M\cos 2\pi\mu_0 x) \tag{10}$$

where t_0 is the bias transmission, M is the modulation, b is the period, and μ_0 is the spatial frequency of the grating. The grating elements run parallel to the y axis. When $M = 1$, the grating has unit contrast and its transmission function lies between 0 and $2t_0$. Let us now take another sinusoidal grating, which is inclined with respect to the first grating. Its transmission function $t_2(x, y)$ is expressed as follows:

$$t_2(x, y) = t_0\left[1 - M\cos\frac{2\pi}{a}(x\cos\theta - y\sin\theta)\right]$$
$$= t_0[1 - M\cos 2\pi(\mu x - \nu y)] \tag{11}$$

The grating is inclined at an angle θ with the y axis and has a period of a. Its spatial frequencies along the x and the y directions are represented by μ and ν, respectively such that $1/a^2 = \mu^2 + \nu^2$. Further, both the gratings are assumed to have the same modulation.

When the gratings are exposed on the same transparency, the transmission function of the positive transparency, with proper processing, can be taken as being proportional to the sum of the two transmission functions: that is,

$$t(x, y) \propto t_1(x) + t_2(x, y) = 2t_0\{1 - M\cos\pi[\mu + \mu_0)x - \nu y]\cos\pi[(\mu - \mu_0)x - \nu y]\} \tag{12}$$

This transmission function corresponds to a grating that is modulated by a low frequency grating (i.e., the moiré pattern). Bright moiré fringes are formed when

$$\cos\pi[(\mu - \mu_0)x - \nu y] = -1$$

or

$$(\mu - \mu_0)x - \nu y = (2m + 1) \tag{13}$$

where m is an integer. Similarly, dark moiré fringes are formed when

$$\cos\pi[(\mu - \mu_0)x - \nu y] = 1 \qquad \text{or} \qquad (\mu - \mu_0)x - \nu y = 2m$$

The moiré fringes are inclined at an angle ϕ with the y axis such that

$$\cot\phi = \frac{\mu - \mu_0}{\nu} = \frac{\cos\theta - a/b}{\sin\theta} \tag{14}$$

When $a = b$, we have

$$\cot\phi = \frac{\cos\theta - 1}{\sin\theta} \tag{15}$$

The period of the moiré pattern is also obtained as follows:

$$d = \left[\frac{1}{(\mu - \mu_0)^2} + \frac{1}{\nu^2}\right]^{1/2} = \frac{ab}{(a^2 + b^2 - 2ab\cos\theta)^{1/2}} \tag{16}$$

These are the same formulas obtained for line gratings.

When the gratings are recorded on separate films and the moiré pattern due to their overlap is observed, the transmission function is obtained by the multiplication of their respective transmission functions, that is, $t(x, y) = t_1(x)t_2(x, y)$. The moiré pattern is then obtained by following the procedure explained earlier.

III. MOIRÉ BETWEEN REFERENCE AND DEFORMED GRATINGS

When the moiré phenomenon is used for metrology, one of the gratings is mounted on the object that is subjected to deformation. Therefore, one then observes the moiré between the deformed grating and the reference grating (undeformed grating). One can also obtain a moiré pattern between two deformed gratings (i.e., when two deformed states of the object are compared). It is thus instructive to study the moiré formation between deformed and undeformed gratings and to learn how to extract the information about the deformation from the moiré pattern.

The deformation is represented by a function $f(x, y)$ which is assumed to be slowly varying. The deformed grating can be expressed as follows:

$$x + f(x, y) = mb \qquad \text{for } m = 0, \pm 1, \pm 2, \pm 3, \pm 4\ldots \tag{17a}$$

This grating is superposed on a reference grating represented by

$$x = nb \qquad \text{for } n = 0, \pm 1, \pm 2, \pm 3, \pm 4, \ldots \tag{17b}$$

This results in a moiré pattern as follows:

$$f(x, y) = pb \qquad \text{for } p = 0, \pm1, \pm2, \pm3, \pm4, \ldots \tag{18}$$

The moiré fringes represent a contour map of $f(x, y)$ with period b. Here, the elements in both the gratings run parallel to the y axis with the deformed grating exhibiting slow variation. We can also obtain moiré between the deformed grating and a reference grating that is oriented at an angle θ with the y axis. That is, the gratings are expressed as just shown:

$$x + f(x, y) = mb \tag{17a}$$

and as

$$y = x \cot\theta - \frac{nb}{\sin\theta} \tag{19}$$

The moiré pattern is given by

$$y + \frac{f(x, y)}{\theta} = p\frac{b}{\theta} \tag{20}$$

This describes a moiré grating with the period and distortion function magnified by a factor $(1/\theta)$.

When two distorted gratings are superposed, the moiré pattern gives the difference between the two distortion functions. This difference can also be magnified when finite fringe mode of moiré formation is used.

The derivative of the distortion function is obtained by observing the moiré between the deformed grating and its displaced replica. For example, let us consider again the deformed grating represented as

$$x + f(x, y) = mb \tag{17a}$$

Its replica has been displaced along the x direction by Δx, hence is represented by

$$x + \Delta x + f(x + \Delta x, y) = nb \tag{21}$$

When these grating are superposed, the moiré pattern is given by

$$\Delta x + f(x + \Delta x, y) - f(x, y) = pb \tag{22}$$

In the limit, when the lateral shift Δx is small, we obtain

$$\Delta x + \frac{\partial f(x, y)}{\partial x}\Delta x = pb \tag{23}$$

The first term is a constant and just represents a shift of the moiré pattern. The moiré pattern thus displays the partial x-derivative of distortion function $f(x, y)$. As mentioned earlier, the moiré effect can be magnified by $(1/\theta)$ using the finite fringe mode.

Pitch mismatch between the gratings can also be used to magnify the effect of distortion. As an example, we consider a distorted grating and a reference grating given by

$$x + f(x, y) = mb \tag{17a}$$

and

$$x = na \tag{24}$$

The moiré pattern is given by

$$x + \frac{a}{|a - b|}f(x, y) = \frac{ab}{|a - b|}p \tag{25}$$

The period of the moiré pattern is $ab/|a - b|$ and the distortion function has been magnified by $a/|a - b|$.

IV. MOIRÉ PATTERN WITH DEFORMED SINUSOIDAL GRATING

The transmission function of a deformed sinusoidal grating is expressed as follows:

$$t_1(x, y) = A_0 + A_1 \cos\frac{2\pi}{b}[x - f(x, y)] \tag{26}$$

where A_0 and A_1 are the constants specifying bias transmission and the modulation of the grating and $f(x, y)$ represents the distortion of the grating. The reference grating, oriented at an angle θ, is represented by

$$t_2(x, y) = B_0 + B_1 \cos\frac{2\pi}{a}(x\cos\theta - y\sin\theta) \tag{27}$$

Here, B_0 and B_1 are constants.

A. Multiplicative Moiré Pattern

Moiré fringes are formed when the angle θ is small, the deformation $f(x, y)$ varies slowly in space, and the periods of the two gratings are nearly equal; in general, $a = Nb$, where N is an integer. The transmission function for multiplicative moiré is the product of their individual transmission functions, that is,

$$\begin{aligned} t(x, y) = t_1(x, y)t_2(x, y) &= A_0B_0 + A_1B_0\cos\frac{2\pi}{b}[x - f(x, y)] \\ &+ A_0B_1\cos\frac{2\pi}{a}[x\cos\theta - y\sin\theta] \\ &+ A_1B_1\cos\frac{2\pi}{b}[x - f(x, y)]\cos\frac{2\pi}{a}[x\cos\theta - y\sin\theta)] \end{aligned} \tag{28}$$

For simplicity, the contrast of both gratings is assumed to be same (i.e., $A_0 = B_0$ and $A_1 = B_1$). Thus we write

$$\begin{aligned} t(x, y) = A_0^2 &+ A_1A_0\cos\frac{2\pi}{b}[x - f(x, y)] \\ &+ A_0A_1\cos\frac{2\pi}{a}[x\cos\theta - y\sin\theta)] \\ &+ \frac{A_1^2}{2}\left\{\cos 2\pi\left\{x\left[\frac{1}{b} + \frac{\cos\theta}{a}\right] - y\frac{\sin\theta}{a} - \frac{f(x, y)}{b}\right\}\right. \\ &\left. + \cos 2\pi\left\{x\left[\frac{1}{b} - \frac{\cos\theta}{a}\right] + y\frac{\sin\theta}{a} - \frac{f(x, y)}{b}\right\}\right\} \end{aligned} \tag{29}$$

In Eq. (29), the first term is a dc term; the second, third, and fourth terms are the carriers; and the fifth term represents the moiré pattern. Moiré fringes are formed wherever

$$x\left[\frac{1}{b}-\frac{\cos\theta}{a}\right]+y\frac{\sin\theta}{a}-\frac{f(x,y)}{b}=p \tag{30}$$

The moiré fringes are deformed straight lines due to $f(x, y)$, the local deformation being $f(x, y)/\theta$, when θ is small.

B. Additive Moiré Pattern

An additive moiré pattern is obtained when the transmission functions of the individual gratings are added. Therefore, the transmission function, assuming gratings of the same modulation, is given by

$$\begin{aligned} t(x,y) = t_1(x,y) + t_2(x,y) = 2A_0 + A_1 \cos\frac{2\pi}{b}[x - f(x,y)] \\ + A_1 \cos\frac{2\pi}{a}[x\cos\theta - y\sin\theta)] \end{aligned} \tag{31}$$

This can be written as follows:

$$\begin{aligned} t(x,y) = 2A_0 + 2A_1 \cos\pi\left\{x\left[\frac{1}{b}+\frac{\cos\theta}{a}\right]\right. \\ \left.-\frac{\sin\theta}{a}-\frac{f(x,y)}{b}\right\}\cos\pi\left\{x\left[\frac{1}{b}-\frac{\cos\theta}{a}\right]+\frac{\sin\theta}{a}-\frac{f(x,y)}{b}\right\} \end{aligned} \tag{32}$$

The second cosine term represents the moiré pattern, which modulates the carrier grating. Cosine variation of the moiré term causes the phase of the carrier to change intrinsically at the crossover point. The visibility of the moiré pattern is generally poor, and imaging optics is required to resolve the carrier grating.

V. CONTRAST IMPROVEMENT OF THE ADDITIVE MOIRÉ PATTERN

It has been mentioned that the contrast of additive moiré pattern is poor. To observe the fringe pattern, it is necessary either to resolve the carrier grating or to use nonlinear recording. Alternately, one could use an FT processor to filter out the required information. The transparency $t(x, y)$ is placed at the input of the FT processor, and the spectrum, which would consist of zero order and two first orders, is observed at the filter plane. Either of the first orders is filtered out and used for imaging. The intensity distribution at the output plane is proportional to

$$\begin{aligned} & A_1^2 \cos^2 \pi \left\{ x \left[\frac{1}{b} - \frac{\cos\theta}{a} \right] + \frac{\sin\theta}{a} - \frac{f(x, y)}{b} \right\} \\ & = \frac{1}{2} A_1^2 \left\{ 1 + \cos 2\pi \left\{ x \left[\frac{1}{b} - \frac{\cos\theta}{a} \right] + \frac{\sin\theta}{a} - \frac{f(x, y)}{b} \right\} \right\} \end{aligned} \tag{33}$$

This represents a unit contrast moiré pattern.

VI. THE MOIRÉ PHENOMENON FOR MEASUREMENT

The moiré phenomenon has been used extensively for the measurement of length and angle. It has also been applied to the study of deformation of objects and also for shape measurement. We will now investigate how to employ this tool in the area of solid mechanics.

A. Measurement of In-Plane Displacement

1. Reference and Measurement Gratings of Equal Pitch and Aligned Parallel to Each Other

A measurement grating is bonded to the object and the reference grating aligned parallel to it. In-plane displacement in the direction of the grating vector (direction normal to the grating elements) causes the period of the bonded grating to change from b to a. The moiré fringes formed will have a period $d = ab/|a - b|$. The normal strain ε as measured by the moiré method is $|a - b|/a$.

Therefore, $\varepsilon = b/d$ (the ratio of the grating period to that of the moiré pattern). At this juncture, it may be noted that the moiré method measures the Lagrangian (engineering) strain. However, if the deformation is small, the Lagrangian and the Eulerian strains are practically equal.

The shear strain is obtained likewise: the measurement grating (grating bonded on the object), as a result of shear, is inclined at an angle θ with the reference grating, leading to the formation of a moiré pattern. Moiré fringes are formed wherever

$$x(1 - \cos\theta) + y\sin\theta = pb \tag{34}$$

The period d of the moiré fringes for very small rotation is b/θ. In fact, the shear strain, when the rotation is small and also when strains are small, is equal to θ. Thus, the shear strain γ is given by

$$\gamma = \frac{b}{d} \tag{35}$$

The shear strain is also obtained as the ratio of the grating pitch to that of the moiré. This is valid for homogeneous normal strain and for simple shear strain.

2. *Two-Dimensional In-Plane Displacement Measurement*

The analysis of the moiré pattern becomes quite easy when it is recognized that the moiré fringes are the contours of the constant displacement component—the so-called isothetics. The zeroth-order moiré fringe runs through regions where the periods of reference grating and measurement grating are equal (i.e., the displacement component is zero). Similarly, the Nth-order moiré fringe runs through the regions where the displacement component is N times the period of the grating. If the reference grating has its grating vector along the x direction, the moiré fringes represent loci of constant u displacement (i.e., $u = Na$). If the v component of the displacement is to be measured, both the reference grating and the measurement grating are aligned with their grating vectors along the y direction. The moiré fringes are now the loci of constant v component, hence $v = N'a$, where N' is the moiré fringe order. To obtain both u and v components simultaneously, a cross-grating may be used. The u and v components are isolated by optical filtering.

Let us consider an example in which the moiré pattern is obtained when both the reference and the measurement gratings are aligned with their grating vectors along the x directions. The moiré pattern then represents the u displacement component. Since strain is a derivative of the displacement, absolute displacement measurement is not necessary. Hence, it is not necessary to assign the absolute fringe order to the moiré fringes, instead, we assign the orders arbitrarily. However, we should keep in mind that increasing orders represent tensile strain, while decreasing orders represent compressive strain. For the sake of analysis, the moiré fringes are assigned orders 8, 7, 6, 5, 4 in the moiré pattern shown in Figure 5. We now wish to obtain strain at point A. Through A, we draw a line parallel to the x axis and a displacement curve (the moiré fringe order vs. the position on the object), as shown in Figure 5: this curve represents the u displacement on the line over the object. We now draw a tangent at the point A. The slope of the tangent at this point gives the normal strain along the x direction. That is,

$$\tan\theta = \frac{\partial N}{\partial x} = \frac{1}{a}\frac{\partial u}{\partial x}$$

or

$$\frac{\partial u}{\partial x} = \varepsilon_x = a\frac{\partial N}{\partial x} \tag{36}$$

In fact, the strain field along this line on the object can be obtained from this curve. If another line parallel to the y axis but passing through A is drawn and the displacement curve plotted as shown in Figure 5, we can obtain $\partial u/\partial y = a(\partial N/\partial y)$ at point A. Similarly, we can obtain $\partial v/\partial y = \varepsilon_y = a(\partial N'/\partial y)$ from the v family of displacement fringes obtained by conducting the experiment with the gratings aligned with their grating vectors parallel to the y axis.

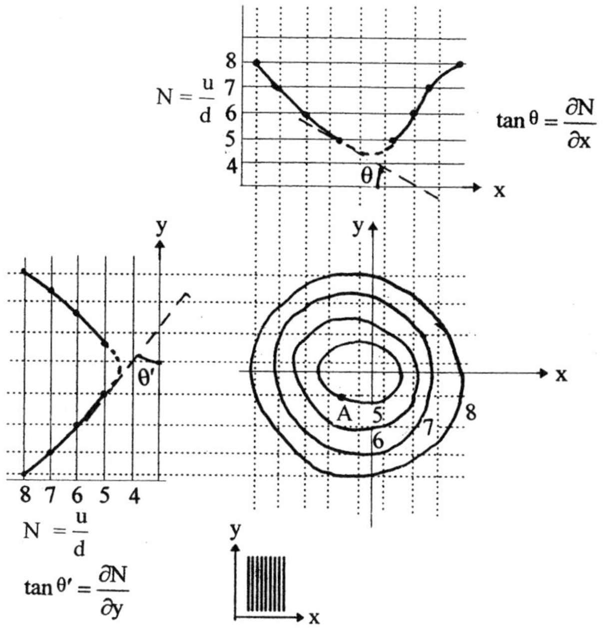

Figure 5 In-plane displacement and normal strain measurement from the moiré.

Also, one obtains the quantity $a(\partial N'/\partial x)$. Thus the shear strain γ_{xy} is calculated from

$$\gamma_{xy} = a\left(\frac{\partial N}{\partial y} + \frac{\partial N'}{\partial x}\right) \tag{37}$$

Therefore, both the normal strains and the shear strain can be obtained with the moiré phenomenon. However, it becomes very clear from this analysis that two factors are very important: assigning the fringe orders to the moiré fringes correctly, generating the displacement curve accurately.

Assignment of the fringe orders is not trivial and requires considerable effort and knowledge. The rules of topography of continuous surfaces govern the order of fringes. Adjacent fringes differ by plus or minus one fringe order except in zones of local fringe maximum or minimum, where they can have equal fringe orders. Local maxima and minima are usually distinguished by closed loops or intersecting fringes; in topography, such contours represent hills and saddlelike features, respectively. Fringes of unequal orders cannot intersect. The fringe order at any point is unique and independent of the path of the fringe count used to reach that point. A fringe can be assigned any number, as absolute displacement information is not required for strain measurement. An arbitrary datum can be used to determine relative displacement.

For generating a displacement curve accurately, one needs a large number of data points. Hence methods have been found to increase the number of moiré fringes, hence the data points, for the same loading conditions. This can be accomplished by pitch mismatch, angular mismatch, or their combination.

The strains can also be obtained by shearing: the moiré patterns representing the displacement fringes are sheared to obtain the fringes corresponding to the strain.

3. High Sensitivity In-Plane Displacement Measurement

The sensitivity of in-plane displacement measurement depends on the period of the grating. In moiré work, low frequency gratings are usually employed and the analysis is based on the geometrical optics. However, with fine-period gratings, though the sensitivity is increased, the diffraction effects are prominent, and consequently quasi-monochromatic, spatially coherent light needs to be used. Indeed, with laser radiation, in-plane displacement measurements can be made with a very high degree of accuracy. A cross-grating, say 1200 l/mm in either direction, recorded holographically and aluminum-coated for reflection, is bonded onto the surface under test. The use of high density gratings puts a limitation on the size of the object that can be examined. The grating is illuminated by four collimated beams as shown in Figure 6. Two beams lie in the x–z plane and two in the y–z plane and make angles such that first-order diffracted beams propagate along the z axis. These beams, on interference, generate fringe patterns characteristic of u and v families of the displacement.

To understand the working of the technique, let us consider a one-dimensional grating with its grating vector along the x direc-

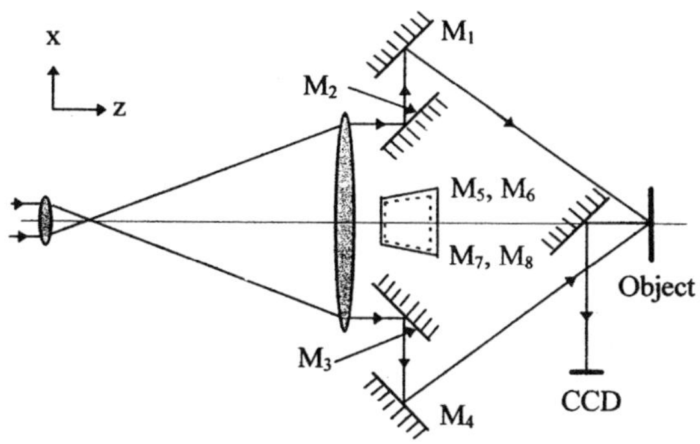

Figure 6 High sensitivity, in-plane measurement setup: mirrors M_5, M_6, M_7, and M_8 generate two beams in the y–z plane.

tion. The grating is recorded holographically by interference between two plane waves, one propagating along the z axis and the other making an angle θ with this but lying in the x–z plane. The period of the grating is b, or its spatial frequency $\mu = \sin\theta/\lambda$ (i.e., $b\sin\theta = \lambda$). This grating is bonded onto the surface of the object. When this grating is illuminated normally by a collimated beam, the first-order diffracted beams make angles of θ and $-\theta$ with the normal to the grating. Let us consider that the object is loaded, resulting in distortions in the grating period. Let the modified spatial frequency be $\mu(x)$. The grating is illuminated symmetrically at angles of θ and $-\theta$ by two plane waves. The plane waves can be represented by $R\exp(2\pi i\mu x)$ and $R\exp(-2\pi i\mu x)$, where R is the amplitude of a wave. These plane waves will be diffracted by the grating and the diffracted field can be expressed as follows:

$$R[\exp(2\pi i\mu x) + \exp(-2\pi i\mu x)]\frac{1}{2}[1 + \cos 2\pi\mu(x)x] \tag{38}$$

Collecting terms of interest (i.e., terms representing waves that propagate along the z axis, we obtain

$$\frac{1}{4}R\{\exp\{2\pi i[\mu - \mu(x)]x\} + \exp\{-2\pi i[\mu - \mu(x)]x\}\} \tag{39}$$

The intensity distribution in the image is

$$I(x) = \frac{1}{4} R^2 \cos^2 2\pi[\mu - \mu(x)]x \tag{40}$$

This represents a moiré pattern. The moiré fringe width is

$$d = \frac{1}{2} \frac{aa(x)}{|a - a(x)|} = \frac{Na}{2} \tag{41}$$

where $a(x)$ is the period of the deformed grating and N is the order of the moiré fringe. Note that the sensitivity of the method is twice as large as would be obtained with the same grating using the conventional method. This is because the deformed grating in the +1 and −1 diffraction orders is being compared. Post has attributed this increase in sensitivity by a factor of 2 to moiré formation between a grating on the test surface and a virtual grating of twice the frequency formed by interference between the two beams, thus providing a multiplicative factor of 2. Use of higher diffraction orders results in increased fringe multiplication.

An arrangement to measure the u and v components of the displacement simultaneously uses a high frequency cross-grating bonded onto the surface of the object. The grating is illuminated simultaneously and symmetrically along the x and y directions to generate four beams traveling axially. The moiré fringes representing the u and v displacement components are then obtained by interference of these four beams. Figure 7 shows the u and v family of fringes for an electronic component that was heated to show the influence of heat (generated internally during operation) on the packaging. A cross-grating with 1200 l/mm is transferred to the face of the electronic component, and He–Ne laser light is used for illumination.

B. Measurement of Out-of-Plane Component and Contouring

The moiré technique is well suited for the measurement of in-plane displacement components: the sensitivity is controlled by the period of the grating. Further, the techniques used for moiré forma-

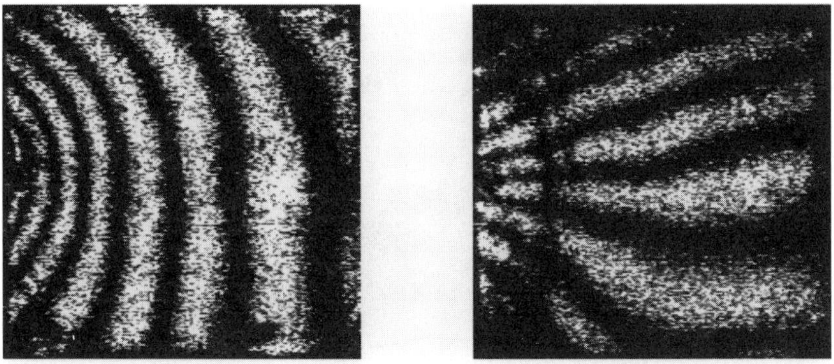

(a) (b)

Figure 7 Interferogram showing (a) u-displacement and (b) v-displacement fringes at the face of an electronic component when energized. (Courtesy of Fu Yu, National University of Singapore.)

tion entail either pitch mismatch or angular mismatch. Therefore, moiré formation, while measuring out-of-plane displacement, will also be governed by these techniques. Consequently, the moiré method for out-of-plane measurement is not as sensitive as that for in-plane measurement. Out-of-plane displacement and surface topography can be measured by one of the following methods: shadow moiré or projection moiré. We will discuss these methods in detail.

C. The Shadow Moiré Method

As the name suggests, a moiré pattern can be formed between the grating and its shadow on the object. Because the shadow grating is distorted by the object topography, one can observe moiré fringes between the distorted and reference gratings.

1. Parallel Illumination and Parallel Observation

Figure 8 shows the arrangement for shadow moiré that relies on parallel illumination and observation; a reference grating is placed on the object. Without loss of generality, we may assume that point A on the object surface is in contact with the grating. The

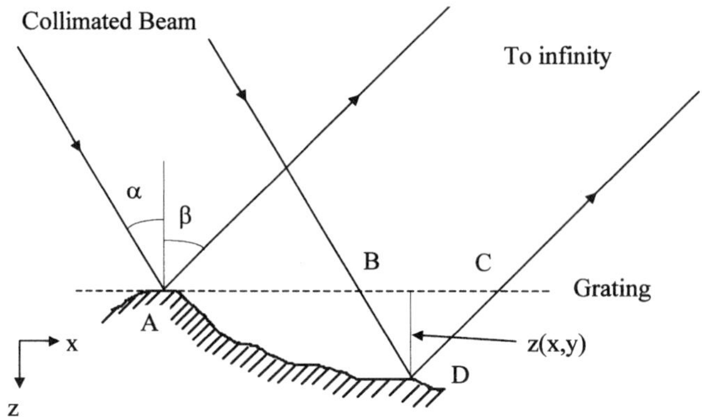

Figure 8 Schematic for shadow moiré.

grating is illuminated by a collimated beam incident at an angle α with the normal to the grating surface (i.e., the z axis). It is viewed from infinity at an angle β. It is obvious that the grating elements contained in a distance AB occupy a distance AD on the object surface. The elements on AD will form a moiré pattern with the grating elements contained in distance AC. Let us assume that AB and AC have p and q grating elements, respectively. So $AB = pa$ and $AC = qa = pb$. From geometry, $BC = AC - AB = (q - p)a = Na$ for $N = 0, \pm 1, \pm 2, \ldots$, N is the order of the moiré fringes. Therefore

$$z(x, y)(\tan \alpha + \tan \beta) = Na$$

or

$$z(x, y) = \frac{Na}{\tan \alpha + \tan \beta} \tag{42}$$

where $z(x, y)$ is the depth as measured from the grating. This is the governing equation of this method. It is seen that the moiré fringes are contours of equal depth measured from the grating. When the viewing is along the normal to the grating (i.e., $\beta = 0$),

$z(x, y) = Na/\tan\alpha$. Alternately, the grating may be illuminated normally and viewed obliquely. Then $z(x, y) = Na/\tan\beta$. This method is not suited for large objects because the assumption that the viewing is done by a collimated beam is not always valid. However, when the object under study is small and the camera is placed sufficiently far away, this requirement is nearly met.

2. Spherical Wave Illumination and Camera at Finite Distance

The assumption made earlier that both the source and the camera are at infinity limits the application of the method to the study of small objects. However, when divergent illumination is used, larger objects can be studied. In general, the source and the camera may be located at different distances from the reference grating. However, a special case (source and camera at equal distances from the grating) is of considerable practical importance and is discussed in detail.

Let the point source S and the camera be at a distance L from the grating surface and their interseparation be P, as shown in Figure 9. The object is illuminated by a divergent wave from the point source. As before, the number of grating elements p con-

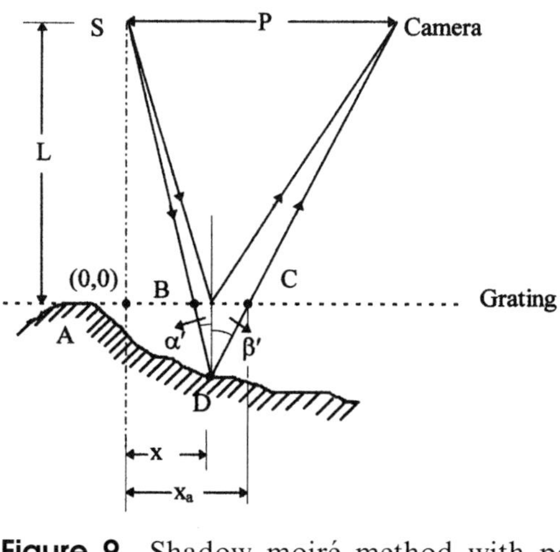

Figure 9 Shadow moiré method with point source illumination and point receiving.

tained in AB on the grating are projected onto AD on the object surface. These elements interact with the elements q in AC, thus producing a moiré pattern. If we assume that Nth-order moiré fringe is observed at the point D, then

$$BC = AC - AB = (q - p)a = Na \tag{43}$$

But $BC = z(x, y)(\tan\alpha' + \tan\beta')$, where $z(x, y)$ is the depth of the point D from the grating. Therefore, we obtain

$$z(x, y) = \frac{Na}{\tan\alpha' + \tan\beta'} \tag{44}$$

Here, α' and β' vary over the surface of the grating or over the surface of the object. From Figure 9 we have $\tan\alpha' = x/[L + z(x, y)]$ and $\tan\beta' = (P - x)/[L + z(x, y)]$. Substituting these in Eq. (44), we obtain

$$z(x, y) = \frac{Na}{\dfrac{x}{L + z} + \dfrac{P - x}{L + z}} = Na\frac{L + z}{P}$$

Rearranging, we obtain

$$z(x, y) = \frac{NaL}{P - Na} = \frac{Na}{\dfrac{P}{L} - \dfrac{Na}{L}} \tag{45}$$

The ratio P/L is called the base-to-height ratio. This is an extremely simple formula. In fact, it is simplicity that makes the technique attractive over the one in which source and camera are placed at different distances from the grating. The distance $\Delta z(x, y)$ between adjacent moiré fringes (i.e., $\Delta N = 1$) is

$$\Delta z(x, y) = \frac{La}{P}\frac{\left(1 + \dfrac{z}{L}\right)^2}{1 - \dfrac{a}{P}\left(1 + \dfrac{z}{L}\right)} \tag{46}$$

It is seen that the fringe spacing is not constant but increases with the depth. Since $a/P \ll 1$, and if $z(x, y)/L \ll 1$ the fringe spacing is constant and is given by $\Delta z(x, y) = La/P$, the moiré fringes represent true depth contours.

Though the method is applicable to the study of large objects, it is essential to correct for the perspective error. This error arises as the actual coordinates (x, y) of a point appear as (x_a, y_a). From Figure 9, the actual coordinates (x, y) and apparent coordinates (x_a, y_a) are related as follows:

$$\frac{x_a - x}{z(x, y)} = \frac{P - x_a}{L}$$

or

$$x = x_a - \frac{z}{L}(P - x_a) = x_a\left(1 + \frac{z}{L}\right) - \frac{zP}{L} \tag{47a}$$

Similarly

$$y = y_a\left(1 + \frac{z}{L}\right) - \frac{zP}{L} \tag{47b}$$

Using this arrangement, the shadow moiré method can be applied to the study, obviously limited by the size of the grating, of very large structures.

Since the fringe spacing is not constant over the object depth, the sensitivity of the method may not be adequate to map out surfaces that are strongly curved or very steep. Therefore, the use of a composite grating has been suggested. The composite grating consists of two parallel superposed gratings with two discretely different periods. This therefore provides two sensitivities.

One of the problems associated with the shadow moiré method is that the moiré fringes do not localize on the surface of the object. Also, the contrast of the fringes is not constant. Fringes of maximum contrast are obtained when the surface under test is nearly flat. The moiré fringes become fuzzy when objects of large curvature are examined. As a caution, objects with large curvature should be placed as close to the grating as possible, and the depth of field of the photographic objective should be large enough to focus the grating and its shadow simultaneously.

D. Automatic Shape Determination

When the distance between the grating and the object surface changes, the moiré fringes shift. From the direction of the shift—which can differ locally—information about the local surface gradient can be obtained. In one of the approaches for automatic surface determination, the grating placed on the object is illuminated obliquely, and a moiré pattern is recorded as shown in Figure 10. A number of moiré patterns are recorded by shifting the grating axially. The phase of the moiré fringes is calculated by means of well-known phase-shifting techniques. The appropriate expression is then used to calculate the object topography.

E. Projection Moiré

There are several differences between the shadow moiré and the projection moiré techniques. In projection moiré, a higher frequency grating is used; the grating is projected on the object surface and is reimaged on a reference grating that is identical to the projection grating and aligned parallel to it. The moiré fringes are formed at the plane of the reference grating. The projection moiré

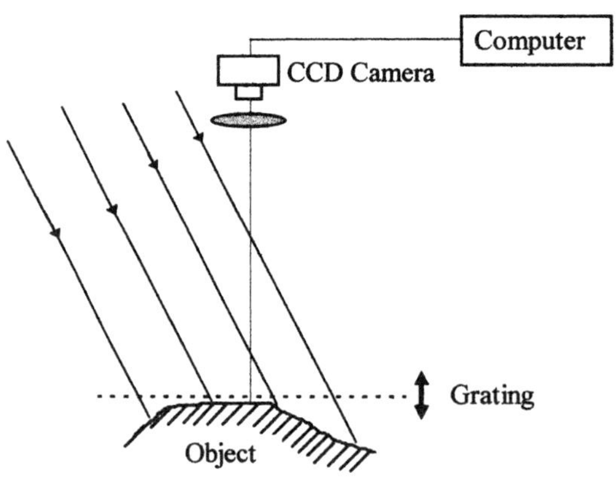

Figure 10 Automatic shape determination.

method also provides a certain flexibility not possible with the shadow moiré method. We will examine the projection moiré method with respect to two cases:

the optical axes of the projection and imaging (observation) systems are parallel
the optical axes are inclined with each other

Following Figure 11 the grating G_1 of period b is imaged on a reference plane where its period is Mb, M being the magnification of the projection system. If the object surface were plane and located on this reference plane, the projected grating would have a constant period. This grating is reimaged on the reference grating G_2. Assuming projection and imaging systems to be identical, the pitch of the imaged grating will be equal to that of G_2 and the grating elements will lie parallel to each other because of the initial alignment. Hence, no moiré pattern will be formed. However, if the surface is curved, the projection grating on the surface will have varying period, hence a moiré pattern will form. We can examine this exactly in the same manner as applied to shadow moiré, that is, by illuminating the grating of period Mb at the reference plane

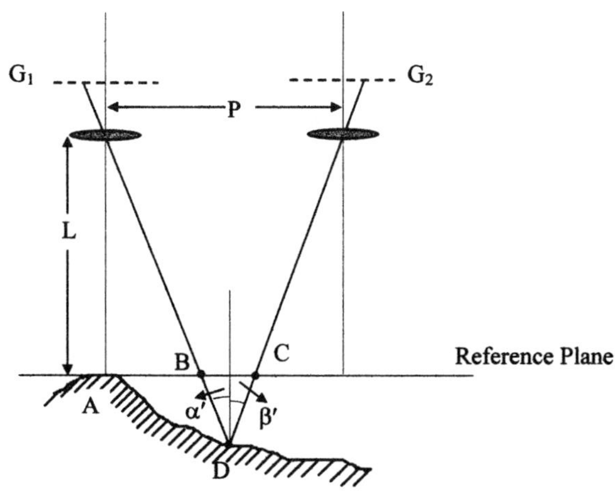

Figure 11 Projection moiré method.

by a spherical wave from the exit pupil of the projection lens. As described in the shadow moiré method, we can write

$$z(x, y) = \frac{NMb}{\tan\alpha' + \tan\beta'} = NMb\frac{L + z(x, y)}{P} \tag{48}$$

Equation (48), on rearranging, becomes

$$z(x, y) = \frac{NMbL}{P - NMb} \tag{49}$$

Writing for the magnification of the imaging system as $M = (L - f)/f$, where f is the focal length, we obtain

$$z(x, y) = \frac{NbL(L - f)}{fP - Nb(L - f)} \tag{50}$$

Here, $z(x, y)$ corresponds to the depth of the object at the point where the Nth-order moiré fringe is formed. It is seen that the moiré fringe spacing is not constant. However, when $Mb \ll P$, the moiré fringes are equally spaced.

The disadvantage of this method is that only very small objects can be studied. However, when the optical axes of the projection system and the imaging system are inclined as shown in Figure 12, large objects can be easily studied. The grating can be imaged on the reference plane, where the Scheimpflug condition is satisfied. The magnification is not constant, however, and consequently the period of the grating on the reference plane varies. This problem can be solved by using a special grating whose period remains constant on projection at the reference plane.

The technique can be used for the comparison of two objects if, instead of the reference grating, a photographic record of the projected grating is made first with one object, and then with the original object replaced by another object. The records may be made on the same film (additive moiré) or on two separate films (multiplicative moiré). Likewise, one can study the out-of-plane deformation of the object by means of recording gratings corresponding to two states of the object.

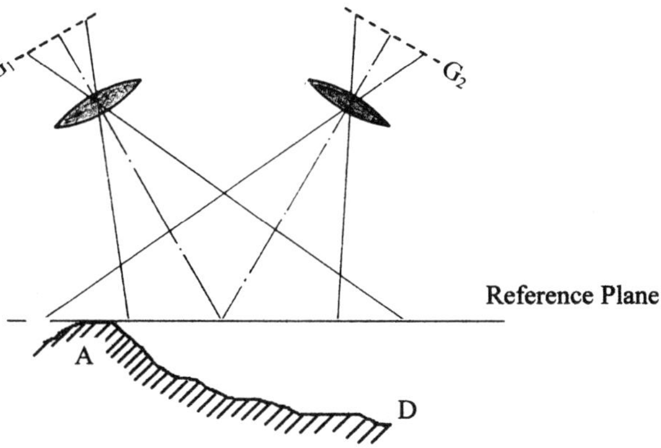

Figure 12 Projection moiré method with oblique incidence and observation.

As may be obvious, the grating plays a very important role: the sensitivity depends on the grating period. Therefore, several gratings may be required to maintain accuracy over steeply curved surface. Further, for phase shifting, the grating needs to be shifted. Use of programmed LCD panels for generating gratings of variable period, sinusoidal or binary transmission function, and arbitrarily inclined, greatly increases the convenience of automatic shape determination.

F. Light Line Projection with TDI Mode of Operation of CCD Camera

If the depth determination of an object is to be carried out over the whole 360°, several measurements are needed. These measurements are made by changing the viewing and the projection directions. An alternative approach is to unwrap the object surface. This is particularly straightforward if the object has axial symmetry. The object is placed on a turntable with its axis coincident with the axis of rotation. A beam of light in the form of a line is projected on the surface as shown in Figure 13a. An image is made on a CCD camera; the image of the projected line will be smeared out as a result of rotational motion. On the other hand, if

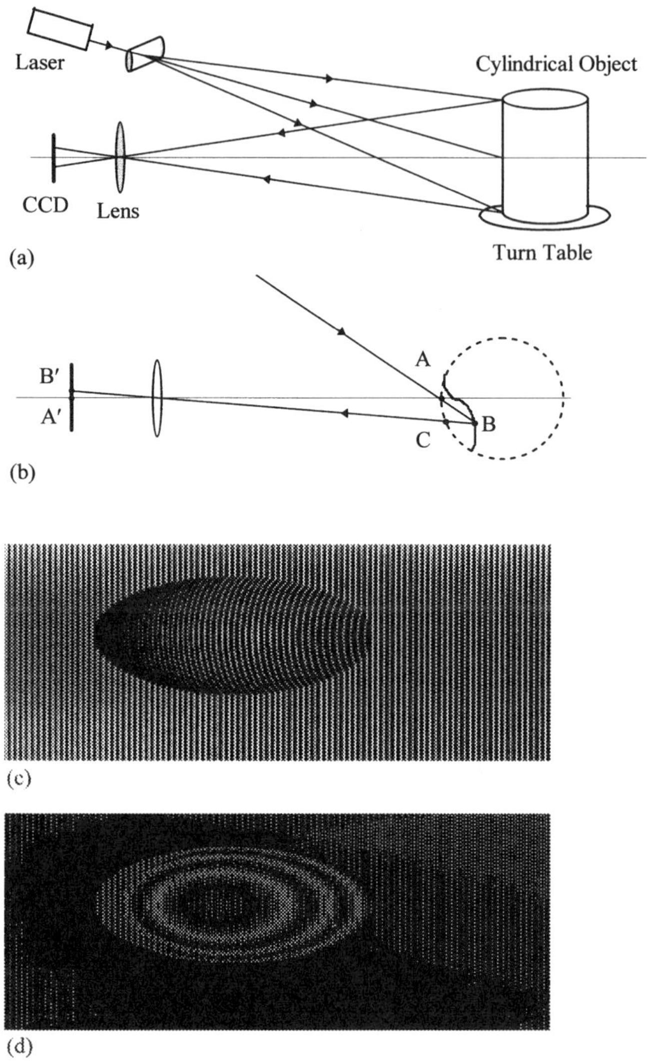

Figure 13 TDI mode of CCD camera: (a) line scan of an object, (b) calculation of the shift of line due to out-of-plane displacement, (c) unwrapped image of a cylindrical object with a dent, and (d) moiré pattern showing the dented region. (Courtesy of Dr. M.R. Sajan.)

the laser is operated in the pulse mode and the camera in TDI (time-delay and integration) mode, the object surface will be unwrapped and will become covered with equidistant lines. The line width will be governed by the width of the projected line and the line separation by the pulse rate. If the object departs from axial symmetry, the pitch of the grating will not be uniform but will vary depending on the magnitude of departure from the nominal cylindrical surface.

Let us consider a section of the object as shown in Figure 13b. It is illuminated at an angle θ with the optical axis. The dashed line is the nominal circular section, while the solid line represents the surface of the object at that section. Point B on the surface is imaged at B', which would also be the image of point C if the object were not deformed. Further, if the object were not deformed, point A would be imaged on axis as A' instead of point B illuminated now. Therefore, the image of the line on the object would be distorted. From the figure, we have $AC = BC \tan\theta$. The out-of-plane displacement d is BC. Thus $d = AC \cot\theta$. Figure 13c shows a distorted image on the unwrapped object. This distortion is seen vividly when a moiré is formed with an internally generated electronic grating as shown in Figure 13d. The moiré condition as before is $AC = Np$, where p is the grating pitch. Therefore, $d = Np \cot\theta$. The out-of-plane displacement is obtained from the moiré pattern. Phase shifting can easily be introduced by the shift of the laser diode.

If the pulse width is larger than the scanning time for one column of detectors, the bright line will be smeared. Hence, the pulse width is determined according to the selected scanning frequency. For example, for a scanning frequency of 50,000 l/min (i.e., scanning time T of 1.2 ms/l), the suggested pulse width should be less than 10% of T (i.e., 0.12 ms). Figure 13c was obtained with a TDI camera having 192 pixels per column (line) and 165 lines. In the TDI mode, the charge of each pixel on a line is added to itself as it moves from one end of the detector to the other. There is thus a time delay from the moment a line of pixels is first illuminated to the instant when it is read into the buffer, and the charge of the pixel is integrated over 165 lines.

G. The Coherent Projection Method

So far we have discussed projection moiré methods in which a grating is projected on the object surface, and its image formed by another optical system on an identical grating. The illumination is incoherent, and the method requires two physical gratings. However, it is possible to use the interference between two coherent waves to create a grating structure on the object. We shall consider two distinct situations: one in which two plane (collimated) waves interfere, generating equidistant, equiphasal plane surfaces in space, and one in which interference between two spherical waves is used to generate interference surfaces (hyperboloidal surfaces).

1. Interference Between Two Collimated Beams

There are several ways of generating two inclined collimated beams. Let us consider two plane waves as shown in Figure 14, which enclose a small angle $\Delta\theta$ between them. Assuming the amplitudes of these waves to be equal, the total amplitude $u(x, y)$ at the x–z plane can be written as follows:

$$u(x, y) = \exp\left\{\frac{2\pi}{\lambda} i\left[x \sin\left(\theta - \frac{\Delta\theta}{2}\right) + z\cos\left(\theta - \frac{\Delta\theta}{2}\right)\right]\right\} + \exp\left\{\frac{2\pi}{\lambda} i\left[x \sin\left(\theta + \frac{\Delta\theta}{2}\right) + z\cos\left(\theta + \frac{\Delta\theta}{2}\right)\right]\right\} \tag{51}$$

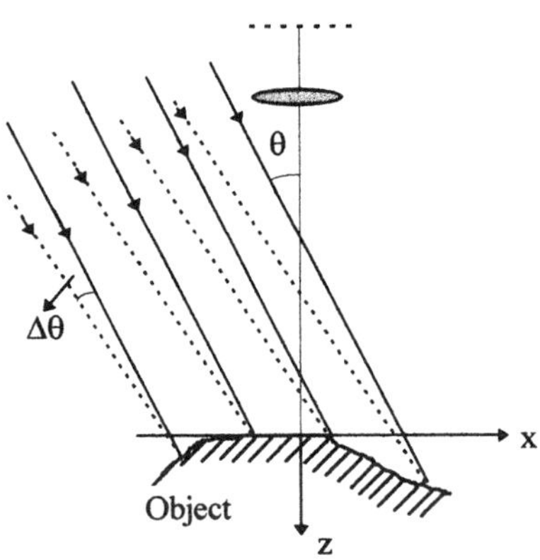

Figure 14 Coherent projection method.

where θ represents the mean direction. When such an amplitude distribution is recorded, the intensity distribution in the record is

$$\begin{aligned} I &= I_0\left[1+\cos\frac{2\pi}{\lambda}2\sin\frac{\Delta\theta}{2}(x\cos\theta+z\sin\theta)\right] \\ &= I_0\left[1+\cos\frac{2\pi}{a}(x\cos\theta+z\sin\theta)\right] \end{aligned} \tag{52}$$

where $a = \lambda/2\sin(\Delta\theta/2)$. Thus a grating of period $a[= \lambda/2\sin(\Delta\theta/2)]$ is formed on the object surface. The period of the grating along the x direction is $a/\cos\theta$, while that along the z direction is $a/\sin\theta$. This grating is photographed normally. If the object surface is a plane lying in the x–y plane, the grating period is $a_x = a/\cos\theta$, and the period in the photographic record is Ma_x, where M is the magnification.

To visualize the moiré formation, we note that in collimated illumination, the fringe width along the x direction remains unchanged irrespective of the depth of the object. However, there is an in-plane shift that is linear with the depth. So, if the object depth changes by Δz, the x shift is $\Delta z\tan\theta$. If this shift is equal to the grating period a_x, a moiré fringe is formed. The moiré fringe interval is thus governed by $a_x = \Delta z\tan\theta = a/\cos\theta$. This leads to $\Delta z = a/\sin\theta$. In other words, a moiré fringe is formed wherever the object depth changes by $a/\sin\theta$. This is thus a neat method of measuring out-of-plane displacement.

We again have the possibilities of either additive moiré or multiplicative moiré. The first exposure is made with the object in its initial state and the second exposure when it is deformed. The moiré pattern gives out-of-plane displacement. Alternately, a reference record can be made of the grating on a reference plane, and the second record made with dual illumination of the object surface. This way, the surface topography can be obtained. The requirement is that the grating period be resolved. Therefore, the interbeam angle should be only a few degrees: this will give a grating structure of about 100 l/mm, which can be resolved by a good imaging system. Obviously, the size of the object that can be studied depends on the size of the collimated beams, which is limited by the collimator optics.

2. Interference Between Two Spherical Waves

To cover large objects, the interference between diverging waves is utilized. One simple method employs a single-mode 50:50 fiber-optic beam splitter (coupler). The laser beam is coupled to one port of the coupler and the beam is equally divided into two output ports. The separation between the output fibers can be adjusted to achieve the desired fringe width. Unfortunately, the surfaces of constant intensity in the interference pattern are not plane surfaces but hyperboloids. The period of the grating on the reference plane thus varies, but it can be calculated if the coordinates of the point sources are known. This method can thus be used with large objects.

H. Measurement of Vibration Amplitudes

Both the shadow moiré and the projection moiré methods can be used to measure the amplitude of vibration. We use the method described earlier in which two plane waves are superposed on the object surface. A record of this will be a grating on the surface of the object with the intensity distribution given by Eq.(52). Let us now assume that the object surface executes a simple harmonic motion of amplitude $w(x, y)$ at frequency ω. We can express this as follows:

$$z = z_0 + w(x, y) \sin \omega t \tag{53}$$

where z_0 represents the static position of the surface. Surface motion will cause the phases of the two interfering beams at the surface to change with time; consequently, the instantaneous intensity distribution in the record will be

$$I(x, t) = I_0 \left\{ 1 + \cos \frac{2\pi}{a} [x \cos \theta + (z_0 + w(x, y) \sin \omega t) \sin \theta] \right\} \tag{54}$$

This intensity distribution is integrated over a period T much larger than the period of vibration $2\pi/\omega$, that is,

$$\begin{aligned} I(x) &= \frac{1}{T_0}\int_0^T I(x,t)dt \\ &= I_0\left\{1+J_0\left[\frac{2\pi}{a}w(x,y)\sin\theta\right]\cos\frac{2\pi}{a}(x\cos\theta+z_0\sin\theta)\right\} \end{aligned} \tag{55}$$

where $J_0(x)$ is the Bessel function of zeroth order. This again represents a grating with a modulation given by the Bessel function. Information about the vibration amplitude $w(x,y)$ can be extracted from this record by optical filtering. We assume that the record has an amplitude transmittance proportional to the intensity distribution. The record is placed at the input plane of a FT processor, and one of the first orders is allowed for image formation. The intensity distribution at the output plane is proportional to

$$J_0^2\left(\frac{2\pi}{a}w(x,y)\sin\theta\right) \tag{56}$$

The intensity distribution exhibits maxima and minina. The amplitude of vibration is obtained as explained for time-averaged holographic interferometry in Chapter 5.

I. The Reflection Moiré Method

Unlike the shadow and projection moiré methods, the reflection moiré method yields information about the slope and the curvature. In the problem of flexure of thin plates, the second derivatives of the deflection are related to the bending moments and twist. The deflection data obtained from shadow or projection moiré need to be differentiated twice to obtain curvature; and to cope with the resulting inaccuracies, one must acquire slope or curvature data so that either a single differentiation is performed or this operation is eliminated altogether. The reflection moiré method serves this purpose. The only disadvantage is that the specimen must be specular.

Figure 15 shows a schematic of the experimental setup for reflection moiré, also known as Ligtenberg's arrangement. The object is an edge-clamped plate. The surface of the plate is mirrorlike, so that a virtual image of the reference grating is formed. Figure 15

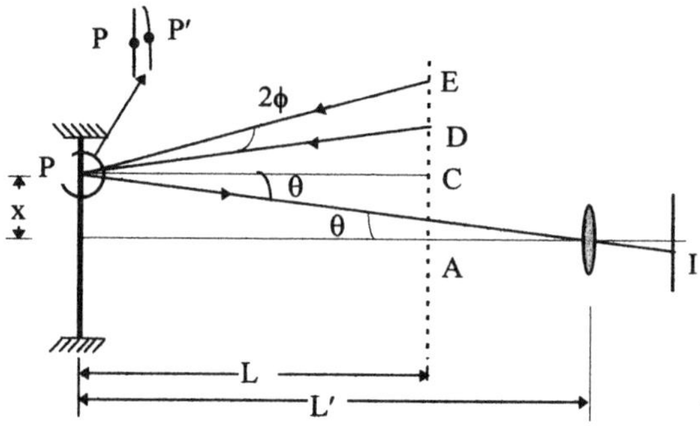

Figure 15 Reflection moiré: Ligtenberg's arrangement.

shows a ray from point D on the grating, which is reflected from point P on the object to point I at the image plane. In other words, point I is the image of point D as formed by reflection on the plate surface. When the plate is deformed, point P moves to point P'. If the local slope is ϕ, point I now receives the ray from point E—and the image of point E is formed on I again. In reality, an image of the grating as seen on reflection from the plate is recorded. This image itself may contain distortions due to nonflatness of the plate surface. It can be shown that self-healing prevents these distortions from influencing moiré formation. After the plate has been deformed (loaded), the image is again recorded on the same photographic film or plate. Loading will further distort the grating image, thus forming a moiré pattern. Following the arguments presented earlier, a grating AE, after loading, is superposed on another grating AD, leading to moiré formation due to pitch mismatch. Let there be p grating elements in AE, and q elements in AD. If the Nth-order moiré fringe appears at point I, then

$$ED = AE - AD = pb - qb = (p - q)b = Nb \tag{57}$$

But

$$\begin{aligned} ED &= EC - CD = L\tan(\theta + 2\phi) - L\tan\theta \\ &= L\frac{\tan\theta + \tan 2\phi}{1 - \tan\theta\tan 2\phi} - L\tan\theta \end{aligned}$$

Assuming ϕ to be very small, which usually is the case, we write $\tan 2\phi \approx 2\phi$. Therefore

$$ED = L\frac{2\phi(1 + \tan^2\theta)}{1 - 2\phi\tan\theta} = Nb \tag{58}$$

Further, $2\phi\tan\theta \ll 1$, and writing the partial x-slope $\phi = \partial w/\partial x$, we have

$$\frac{\partial w}{\partial x} = \frac{Nb}{2L(1 + \tan^2\theta)} = \frac{Nb}{2L\left(1 + \dfrac{x^2}{L'^2}\right)} \tag{59}$$

The slope depends on the value of x. To eliminate the dependence of $\partial w/\partial x$ on x, a curved grating is used. However, Ligtenberg's method utilizing curved gratings also suffers from several disadvantages, including the large dimensions of the curved surface gratings and the need to use gratings of low spatial frequency to obtain moiré patterns of good contrast. In addition, applicability is limited to static loading problems and to models of relatively large flexures.

The disadvantages of the Ligtenberg's method are largely removed by the Rieder–Ritter arrangement. A cross-grating is used, so that $\partial w/\partial x$ and $\partial w/\partial y$ are simultaneously recorded. These can be separated by optical filtering.

Further, as with other moiré methods, the sensitivity is dependent on the grating period. It is therefore desirable to have an arrangement in which the grating period can be varied easily. Figure 16 shows a grating that is projected onto a ground-glass screen. The projected grating is imaged using the Rieder–Ritter arrangement. The pitch of the grating can be varied by changing the magnification during projection of the grating.

Several other arrangements have been suggested to obtain slope and curvature. For example, a diffraction grating may be placed at or near the focal plane of the imaging lens to produce sheared fields. Double exposure, before and after the plate loading, is performed. In another interesting arrangement, an inclined plane parallel plate is placed near the focal plane of the imaging lens. The

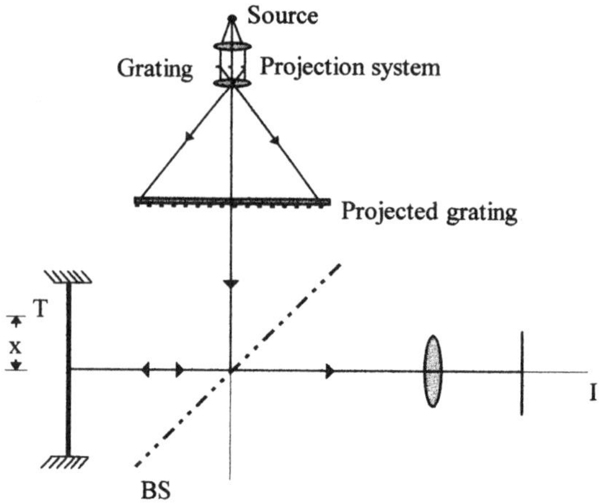

Figure 16 Reflection moiré: grating with variable pitch.

image is recorded in reflection: two reflected beams, which are sheared, produce a grating in the image plane. Loading of the plate deforms this grating. Therefore, double-exposure recording, before and after plate loading, results in the formation of a moiré pattern that is due to the slope variation. When two plates are used in tandem and properly aligned such that three sheared beams participate in interference, curvature fringes are obtained.

These methods are coherent methods requiring laser radiation. The permissible size of the object depends on the collimating optics; hence these methods are limited to objects that are small to moderate in size. As mentioned earlier, the surface of the plate (object) need not be optically flat. The departure from flatness causes some distortions in the grating. But the moiré phenomenon has a self-healing property whereby these distortions cancel out, and the moiré fringes are due solely to the deformation of the plate caused by loading.

VII. SLOPE DETERMINATION FOR DYNAMIC EVENTS

So far we have described methods that are applicable only to static problems. Pulse illumination, in which the grating images corresponding to successive states of the object are recorded, is frequently used for the examination of dynamic processes. Such methods are rather troublesome to apply to the study of dynamic events, however, since to generate the moiré pattern, the reference grating must be recorded each time, along with the image.

Several methods have been developed whereby the images of the reference grating and of the deformed grating from the surface under investigation are recorded simultaneously. In one of these arrangements (Fig. 17), the beam splitter provides two paths for the imaging. One image is formed by reflection from the object surface *T*, while the other image is formed by reflection from a mirror surface *M*. Therefore, both images of the grating can be recorded in a single exposure (pulse illumination), hence an additive moiré pattern can be observed for the various states of the object. This arrangement, however, does not exploit the healing property of the moiré phenomenon. Since the object surface in its initial state and the mirror surface are not identical in general, the grating formed by reflection from the object surface will carry some distortions, which would show up as moiré fringes even for the unloaded state of the object. In other words, the initial fringe-free field is not obtained. This problem is solved in the arrangement shown in Figure 18. The grating G_1 is a replica of the image formed by reflection from the object surface. Therefore, the images

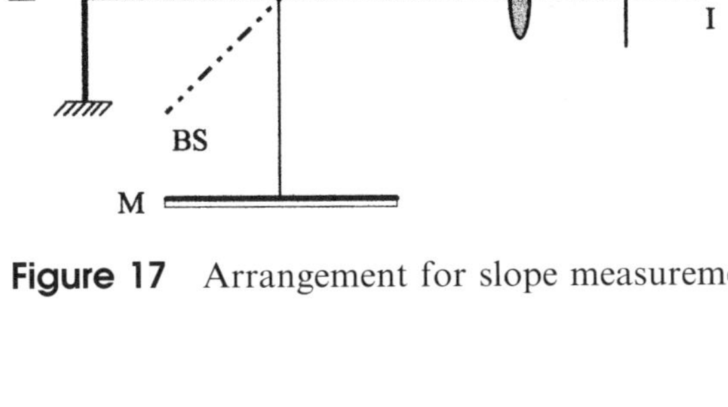

Figure 17 Arrangement for slope measurement.

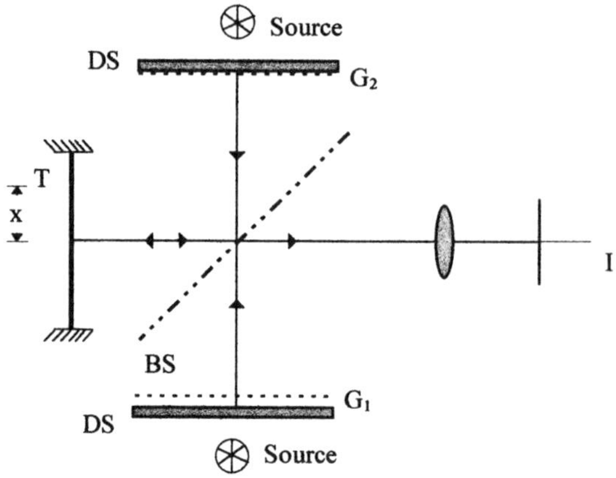

Figure 18 Modified arrangement for slope measurement.

of gratings G_1 and G_2 will be identical and an initial fringe-free field is obtained. However, for each object surface, a new grating G_1 must be produced and aligned.

VIII. CURVATURE DETERMINATION FOR DYNAMIC EVENTS

The arrangements shown in Figures 17 and 18 can be modified to provide for additional shear by incorporating a Mach–Zehnder interferometer. Figure 19 shows this arrangement. The images are sheared by the tilt given to one of the mirrors.

IX. SURFACE TOPOGRAPHY WITH THE REFLECTION MOIRÉ METHOD

The methods described for slope and curvature measurements exploit the imaging of a reference grating when the separation between the grating and the object surface is large. However, when the reference grating is placed very close to the object surface, the moiré pattern is formed between the reference grating and its image. Under certain conditions, the moiré fringes are loci of constant depth. We consider the arrangement shown in Figure 20

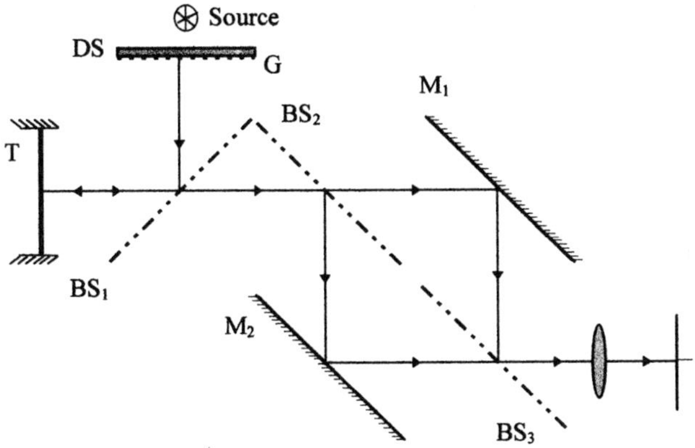

Figure 19 Arrangement for curvature measurement.

to explain the formation of the moiré pattern. Grating G_1 (which essentially would be an image of G_2) is viewed through grating G_2. Let the gratings G_1 and G_2 be described by

$$x = mb \qquad \text{and} \qquad x = na$$

where m and n are integers and the gratings have different periods b and a, respectively. The camera is located at the position (x_c, z_c). The x coordinate of any point projected from the grating G_1 to grating G_2 is obtained from

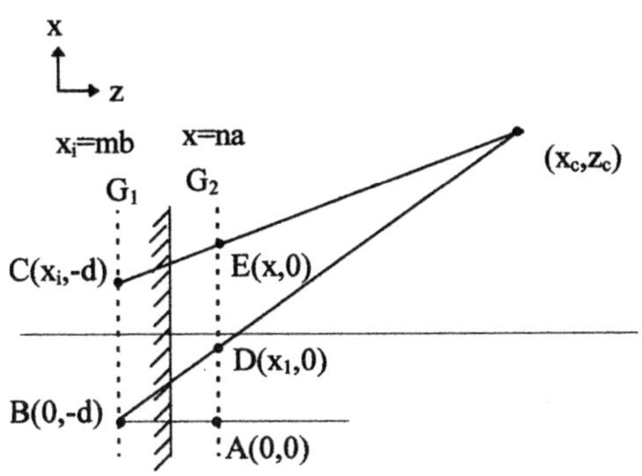

Figure 20 Surface topography via reflection moiré.

$$\frac{x_c - x}{z_c} = \frac{x_c - mb}{z_c + d}$$

or

$$x = \frac{bz_c}{z_c + d}m + \frac{d}{z_c + d}x_c \tag{60}$$

Using indicial equation $m - n = p$, we obtain an equation of the moiré pattern as follows:

$$x = \frac{abz_c}{(a-b)z_c + ad}p + \frac{ad}{(a-b)z_c + ad}x_c \tag{61}$$

In general, d is a function of x and y [i.e., $d(x, y)$]. Let us now consider another experimental arrangement (Fig. 21): a grating is placed on a reflecting surface, and an image of the grating is formed by reflection. Imaging will cause both the period and the separation of the image grating to vary spatially. The moiré pattern is observed by interaction of the image grating with the refer-

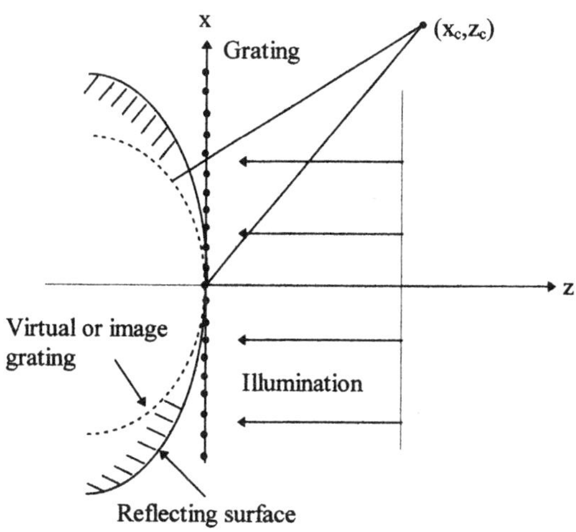

Figure 21 Reflection moiré for contouring.

ence grating. We now assume that the reflecting surface departs very little from the plane surface and that the reference grating is in contact with the surface. Then it is safe to assume that the periods of both the reference and the image gratings are equal (i.e., $a = b$). The moiré pattern is then described by

$$x = \frac{az_c}{d(x, y)} p + x_c \tag{62}$$

Since $d(x, y)$ is the separation between the two gratings, the depth $z(x, y)$ of the surface from the reference grating will be half of it [i.e., $z(x, y) = d(x, y)/2$]. Therefore, we write

$$z(x, y) = \frac{az_c}{2(x - x_c)} p \tag{63}$$

Further, if the area to be examined is very small and the illumination source is far away obliquely, then

$$z(x, y) = -\frac{az_c}{2x_c} p \tag{64}$$

The moiré fringes now represent true depth contours.

We can apply this technique to study moiré pattern formation when the grating is placed on a spherical surface of radius curvature R. Let us assume that the reference grating is placed at $z = 0$. The equation of the spherical surface is

$$x^2 + y^2 + z^2 + 2zR = 0 \tag{65}$$

The distance $z(x, y)$ between the grating and the surface for large R is $z = [(x^2 + y^2)/2R]$.

On substituting for $z(x, y)$, we now write the equation of the moiré pattern as follows:

$$x^2 + y^2 = \frac{az_c R}{x_c} p \tag{66}$$

The moiré fringes are circles of radii $[(az_c R/x_c)p]^{1/2}$. The radii vary as $\sqrt{p}$, as in a Fresnel zone plate. Figure 22 shows photographs of

(a) (b)

Figure 22 Contour maps of a spherical surface with gratings of (a) 200 l/in. and (b) 500 l/in.

such moiré patterns taken by placing gratings of 200 and 500 l/in. on a concave mirror surface.

X. THE TALBOT PHENOMENON

When a periodic object is illuminated by a coherent monochromatic beam, its image is formed at specific planes called self-image planes or Talbot planes. The effect was first observed by Talbot in 1836 and its theory was worked out by Rayleigh in 1881. Self-imaging is due to diffraction and can be observed with periodic objects that satisfy the Montgomery condition. A linear (1D) grating is one such object. For a 1D grating of spatial frequency μ, illuminated by a collimated beam of wavelength λ, the self-image planes are equidistant and are located at distances $N/\mu^2\lambda$ from the object, where $N = 1, 2, 3, \ldots$ gives the order of the Talbot planes. In other words, the transverse periodicity of the object manifests itself as longitudinal periodicity. The imaging is called self-imaging because no imaging devices are used. A two-dimensional (cross-) grating with the same spatial frequency μ in both directions also self-images at the planes located at $N/\mu^2\lambda$ from the grating.

A. The Talbot Effect in Collimated Illumination

To explain Talbot imaging, let us consider a one-dimensional grating whose transmittance is given by

$$t(x) = \frac{1}{2}[1 + \cos 2\pi\mu x]$$

This grating is placed at the $z = 0$ plane and is illuminated by a collimated beam of amplitude A, as shown in Figure 23. The amplitude of the wave just behind the grating ($z = 0$ plane) will be given by

$$u(x, z = 0) = \frac{1}{2}A[1 + \cos 2\pi\mu x] \tag{67}$$

When the Fresnel diffraction approach is used, the amplitude at any plane z is obtained as follows:

$$u(x_1, z) = \frac{A}{2}\exp(ikz)[1 + \exp(-i\pi\mu^2\lambda z)\cos 2\pi\mu x_1] \tag{68}$$

The amplitude distribution at any plane z will be identical to the grating transmittance function except for a constant multiplicative phase factor, if $\exp(-i\pi\mu^2\lambda z) = 1$. This condition is satisfied when

$$\pi\mu^2\lambda = 2N\pi \qquad \text{for } N = 0, 1, 2, 3, \ldots \tag{69}$$

The planes at which this condition is satisfied are the Talbot planes. However, when N takes half-integer values, we still obtain the transmittance function of a sinusoidal grating but it is phase-shifted by π. Thus, the Talbot planes are separated by $z_T = 1/\mu^2\lambda$. In the case of collimated illumination, the Talbot planes are equi-spaced.

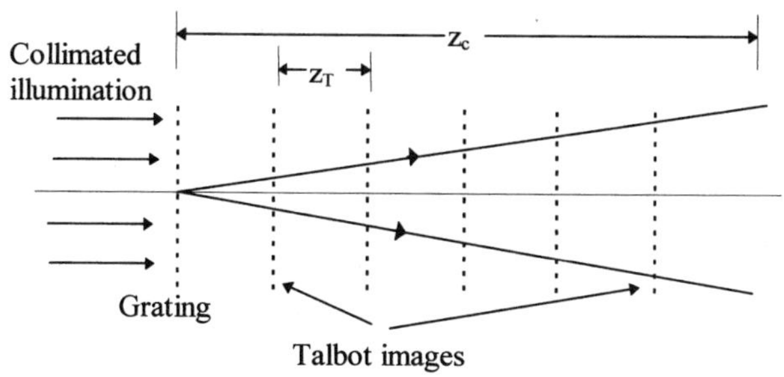

Figure 23 Formation of Talbot images in collimated illumination.

B. Cutoff Distance

The Talbot images are formed as a result of constructive interference among the diffracted waves at successive Talbot planes. For an infinite grating illuminated by an infinitely large beam, the various diffracted waves will continue overlapping irrespective of the distance, thus producing an infinity of Talbot images. In a practical situation, both the grating and the beam are of finite dimensions. Therefore, the diffracted waves will not overlap after a certain distance; consequently no Talbot images are formed after this distance. Let us consider a grating of linear dimension D and spatial frequency μ illuminated by a beam of size D. The cutoff distance z_c is defined as the distance from the grating over which the first-order diffracted beam deviates by half the beam size. This is given by $z_c = D/2\mu\lambda$.

C. The Talbot Effect in Noncollimated Illumination

Let us consider that a point source is placed at (0, 0) and the grating is at the $z = R$ plane. The grating is illuminated by a divergent spherical wave from the point source. The amplitude of the wave just behind the grating is

$$u(x, R) = \frac{A}{2R}\exp(ikR)\exp\left(i\frac{k}{2R}x^2\right)[1 + \cos 2\pi\mu x] \tag{70}$$

This is valid under paraxial approximation. Using Fresnel diffraction approximation, the amplitude at a plane distant z from the grating is

$$\begin{aligned} u(x_1, R + z) = {} & \frac{A}{2(R+z)}\exp[ik(R+z)] \\ & \exp\left[i\frac{k}{2(R+z)}x_1^2\right]\left[1 + \exp\left(-i\pi\frac{Rz}{R+z}\mu^2\lambda\right)\right. \\ & \left.\cos 2\pi\mu\frac{R}{R+z}x_1\right] \end{aligned} \tag{71}$$

This expression represents a transmittance function of the grating of spatial frequency $\mu' = \mu R/(R + z)$, provided

$$\exp\left(-i\pi \frac{Rz}{R+z}\mu^2\lambda\right) = 1$$

This yields the self-image planes distances $z(_{\mathrm{T}})_{\mathrm{s}}$ as follows:

$$(z_{\mathrm{T}})_{\mathrm{s}} = \frac{2N}{\mu^2\lambda - 2N/R} \qquad \text{for } N = 1, 2, 3, \ldots \tag{72}$$

The spacing between the successive Talbot planes increases with the order N. The period of the grating also increases as if it were geometrically projected. Similarly, when the grating is illuminated by a convergent spherical wave, the successive Talbot planes come closer and the spatial frequency increases. This is valid until some distance from the point of convergence.

D. The Talbot Effect for Measurement

Instead of using a projection system for projecting a grating on the object surface for either shadow moiré or projection moiré work, the projection may be done without a lens but using the Talbot effect. This, however, necessitates the use of laser radiation. An additive moiré method can then be applied to measure out-of-plane deformation. The shape of the object can also be obtained by imaging the Talbot grating on another identical grating.

When a grating of higher frequency is used, several Talbot planes may intersect the object surface. A moiré pattern of high contrast will be formed at the Talbot planes. Therefore, this technique can be used for the topographical examination of objects of large depth.

BIBLIOGRAPHY

1. K. J. Gasvik, *Optical Metrology*, John Wiley, Chichester (1987).
2. O. D. D. Soares (Ed.), *Optical Metrology*, Martinus Nijhoff, Dordrecht (1987).
3. Alexis Lagarde (Ed.), *Optical Methods in Mechanics of Solids*, Society for Experimental Mechanics, Inc., Bethel, CT (1989).

4. James F. Doyle and James W. Phillips (Eds.), *Manual of Experimental Stress Analysis*, Society for Experimental Stress Analysis, Brookfield Center, CT (1989).
5. O. Kafri and I. Glatt, *The Physics of Moiré Metrology*, John Wiley, New York (1990).
6. G. Indebetouw and R. Czarnek (Eds.), *Selected Papers on Optical Moiré and Applications*, MS64, SPIE Press, Bellingham, WA (1992).
7. K. Patorski, *Handbook of the Moiré Fringe Technique*, Elsevier, Amsterdam (1993).
8. Daniel Post, Bongtae Han, and Peter Ifju, *High Sensitivity Moiré*, Springer-Verlag, New York (1994).
9. Gary L. Cloud, *Optical Methods of Engineering Analysis*, Cambridge University Press, Cambridge (1995).
10. Nils H. Abramson, *Light in Flight or The Holodiagram; The Columbi Egg of Optics*, Vol. PM 27, SPIE Press, Bellingham, WA (1997).

Appendix: Additional Reading

Phase Evaluation Methods

1. D. A. Tichenor and V. P. Masden, Computer analysis of holographic interferograms for non-destructive testing, *Opt. Eng.*, *8*, 469–473 (1979).
2. T. M. Kreis and H. Kreitlow, Quantitative evaluation of holographic interferograms under image processing aspects, *Proc. SPIE, 210*, 196–202 (1979).
3. M. Takeda, H. Ina and S. Kobayashi, Fourier-transform method of fringe pattern analysis for computer based topography and interferometry, *J. Opt. Soc. Amer.*, *72*, 156–160 (1982).
4. T. Yatagai and M. Idesawa, Automatic fringe analysis for moiré topography, *Opt. Lasers in Eng.*, *3*, 73–83 (1982).
5. P. Hariharan, B. F. Oreb and N. Brown. A digital phase-measurement system for real-time holographic interferometry, *Opt. Commun.*, *41*, 393–396 (1982).
6. H. E. Cline, W. E. Lorensen and A. S. Holik, Automatic moiré contouring, *Appl. Opt.*, *23*, 1454–1459 (1984).
7. V. Srinivasan, H. C. Liu and M. Halioua, Automated phase measuring profilometry of 3-D diffuse objects, *Appl. Opt.*, *23*, 3105–3108 (1984).
8. K. Creath, Phase-shifting speckle interferometry, *Appl. Opt.*, *24*, 3053–3058 (1985).

9. K. A. Stetson and W. R. Brohinsky, Electrooptic holography and its applications to hologram interferometry, *Appl. Opt.*, *24*, 3631–3637 (1985).
10. K. Andresen. The phase shift method applied to moiré image processing, *Optik, 72*, 115–119 (1986).
11. G. T. Reid, R. C. Rixon, S. J. Marshall and H. Stewart, Automatic on-line measurements of 3-D shape by shadow casting moiré topography, *Wear, 109*, 297–304 (1986).
12. K. Andresen and D. Klassen, The phase shift method applied to cross grating moiré measurement, *Opt. Lasers in Eng.*, *7*, 101–114 (1986/87).
13. M. Owner-Petersen and P. Damgaard Jensen, Computer-aided electronic speckle pattern interferometry (ESPI): deformation analysis by fringe manipulation, *NDT Int.*, *21*, 422–426 (1988).
14. K. Creath, Phase-measurement interferometry techniques, in *Progress in Optics, 26*, pp. 349–393 (ed. Emil Wolf), North Holland, Amsterdam, 1988.
15. Y. Morimoto, Y. Seguchi and T. Higashi, Application of moiré analysis of strain using Fourier transform, *Opt Eng.*, *27*, 650–656 (1988).
16. M. Kujawińska and D. W. Robinson, Multichannel phase-stepped holographic interferometry, *Appl. Opt.*, *27*, 312–320 (1988).
17. J. J. J. Dirckx and W. F. Decraemer, Phase shift moiré apparatus for automatic 3D surface measurement, *Rev. Sci. Instrum.*, *60*, 3698–3701 (1989).
18. J. M. Huntley, Noise-immune phase unwrapping algorithm, *Appl. Opt.*, *28*, 3268–3270 (1989).
19. M. Takeda, Spatial-carrier fringe pattern analysis and its applications to precision interferometry and profilometry: An overview, *Industrial Metrology, 1*, 79–99 (1990).
20. M. Kujawińska, L. Salbut and K. Patorski, 3-Channel phase-stepped system for moiré interferometry, *Appl. Opt.*, *30*, 1633–1637 (1991).
21. J. Kato, I. Yamaguchi and S. Kuwashima, Real-time fringe analysis based on electronic moiré and its applications in *Fringe '93* (eds. W. Jüptner and W. Osten), pp. 66–71, Akademie Verlag (1993).

Detectors and Recording Materials

1. J. C. Urbach and R. W. Meier, Thermoplastic xerographic holography, *Appl. Opt.*, *5*, 666–667 (1966).
2. T. A. Shankoff, Phase holograms in dichromated gelatin, *Appl. Opt.*, *7*, 2101–2105 (1968).
3. M. J. Beesley and T. G. Castledine, The use of photoresist as a holographic recording medium, *Appl. Opt.*, *9*, 2720–2724 (1970).
4. W. S. Colburn and K. A. Haines, Volume hologram formation in photopolymer materials, *Appl. Opt.*, *10*, 1636–1641 (1971).
5. B. Smolinska, Relief hologram formation and replication in hardened dichromated PVA films, *Acta Physica Polonica, A40*, 327–332 (1971).
6. T. L. Credelle and F. W. Spong, Thermoplastic media for holographic recording, *RCA Review, 33*, 206–226 (1972).
7. E. G. Ramberg, Holographic information storage, *RCA Review, 33*, 5–53 (1972).
8. D. Meyerhofer, Phase-holograms in dichromated gelatin, *RCA Review, 33*, 110–130 (1972).
9. R. A. Bartolini, Characteristics of relief holograms recorded on photoresists, *Appl. Opt. 13*, 129–139 (1974).
10. A. Graube, Advances in bleaching methods for photographically recorded holograms, *Appl. Opt.*, *13*, 2942–2946 (1974).
11. K. Biedermann, Information storage materials for holography and optical data storage, *Opt. Acta, 22*, 103–124 (1975).
12. B. L. Booth, Photopolymer materials for holography, *Appl. Opt.*, *14*, 593–601 (1975).
13. S. L. Norman and M. P. Singh, Spectral sensitivity and linearity of Shipley AZ 1350 J photoresist, *Appl. Opt.*, *14*, 818–820 (1975).
14. S. K. Case and R. Alferness, Index modulation and spatial harmonic generation in dichromated gelatin films, *Appl. Phys.*, *10*, 41–51 (1976).
15. R. A. Bartolini, H. A. Weakliem and B. F. Williams, Review and analysis of optical recording media, *Ferroelectrics, 11*, 393–396 (1976).
16. R. A. Bartolini, Optical recording media review, *Proc. SPIE, 123*, 2–9 (1977).
17. B. L. Booth, Photopolymer laser recording materials, *J. Appl. Photographic Eng.*, *3*, 24–30 (1977).
18. D. Casasent and F. Caimi, Photodichroic crystals for coherent optical data processing, *Opt. Laser Technol.*, *9*, 63–68 (1977).

19. T. C. Lee, J. W. Lin and O. N. Tufte, Thermoplastic photoconductor for optical recording and storage, *Proc. SPIE, 123*, 74–77 (1977).
20. B. J. Chang, Dichromated gelatin as holographic storage medium, *Proc. SPIE, 177*, 71–81 (1979).
21. K. Blotekjaer, Limitations on holographic storage capacity of photochromic and photorefractive media, *Appl. Opt.*, *18*, 57–67 (1979).
22. B. J. Chang and C. D. Leonard, Dichromated gelatin for the fabrication of holographic elements, *Appl. Opt.*, *18*, 2407–2417 (1979).
23. T. Kubota and T. Ose, Methods of increasing the sensitivity of methylene-blue sensitized dichromated gelatin, *Appl. Opt.*, *18*, 2538–2539 (1979).
24. P. Hariharan, Holographic recording materials: Recent development, *Opt. Eng.*, *19*, 636–641 (1980).
25. P. Hariharan, Silver halide sensitised gelatin holograms: mechanism of hologram formation, *Appl. Opt.*, *25*, 2040–2042 (1986).
26. S. Calixto, Dry polymer for holographic recording materials, *Appl. Opt.*, *26*, 3904–3910 (1987).
27. R. T. Ingwall, M. Troll and W. T. Vetterling, Properties of reflection holograms recorded on Polaroid's DMP-128 photopolymer, *Proc. SPIE, 747*, 67–73 (1987).
28. N. V. Kukhtarev and V. V. Mauravev, Dynamic holographic interferometry in photorefractive crystals, *Opt. Spectrosc. (USSR), 64*, 656–659 (1988).
29. R. Changkakoti and S. V. Pappu, Methylene blue sensitised DCG holograms: a study of their storage and reprocessibility, *Appl. Opt.*, *28*, 340–344 (1989).
30. R. A. Lessard and J. J. Couture, Holographic recordings in dye/polymer systems for engineering applications, *Proc. SPIE, 1183*, 75–89 (1989).
31. R. T. Ingwall and M. Troll, Mechanism of hologram formation in DMP-128 photopolymer, *Opt. Eng.*, *28*, 586–591 (1989).
32. S. Redfield and L. Hesselink, Data storage in photorefractives revisited, *Opt. Comput.*, *63*, 35–45 (1989).
33. W. K. Smothers, B. M. Monroe, A. M. Weber and D. E. Keys, Photopolymers for holography, *Proc. SPIE, 1212*, 20–29 (1990).
34. F. Ledoyen, P. Bouchard, D. Hennequin and M. Cormier, Physical Model of a liquid thin film: application to infrared holographic recording, *Phys. Rev. A, 41*, 4895–4902 (1990).

35. G. Savant and T. Jonnson, Optical recording materials, *Proc. SPIE, 1461*, 79–90 (1991).
36. S. A. Zager and S. M. Weber, Display holograms in du Pont's Omnidex ™ films. *Proc. SPIE, 1461*, 58–67 (1991).
37. F. P. Shvartsman, Dry photopolymer embossing: Novel photoreplication technology for surface relief holographic optical elements. *Proc. SPIE, 1507*, 383–391 (1991).
38. E. Bruzzone and F. Mangili, Calibration of a CCD camera on a hybrid coordinate measuring machine for industrial metrology, *Proc. SPIE, 1526*, 96–112 (1991).
39. J. L. Salter and M. F. Loeffler, Comparison of dichromated gelatin and du Pont HRF-700 photopolymer as media for holographic notch filters, *Proc. SPIE, 1555*, 268–278 (1991).
40. R. D. Rallison, Control of DCG and non silver halide emulsions for Lippmann photography and holography, *Proc. SPIE, 1600*, 26–37 (1991).
41. T. J. Cvetkovich, Holography in photoresist materials, *Proc. SPIE, 1600*, 60–70 (1991).
42. K. Kurtis and D. Psaltis, Recording of multiple holograms in photopolymer films, *Appl. Opt., 31*, 7425–7428 (1992).
43. R. A. Lessard, C. Malouin, R. Changkakoti and G. Mannivannan, Dye-doped polyvinyl alcohol recording materials for holography and non-linear optics, *Opt. Eng., 32*, 665–670 (1993).
44. U. S. Rhee, H. J. Caulfield, J. Shamir, C. S. Vikram and M. M. Mirsalehi, Characteristics of the du Pont photoplymer for angularly multiplexed page-oriented holographic memories, *Opt. Eng., 32*, 1839–1847 (1993).
45. D. Dirksen and G. von Bally, Holographic double-exposure interferometry in near real time with photorefractive crystals, *J. Opt. Soc. Amer. B, 11*, 1858–1863 (1994).
46. K. Meerholz, B. L. Volodin, Sandalphon, B. Kippelen and N. Peyghambarlan, A photorefractive polymer with high optical gain and diffraction efficiency near 100%, *Nature, 373*, 497–500 (1994).
47. Jean-Pierre Fouassler and F. Morlet-Savary, Photopolymers for laser imaging and holographic recording: design and reactivity of photosensitizers, *Opt. Eng., 35*, 304–312 (1996).
48. V. V. Vlad, D. Malacara-Hernandez and A. Petris, Real-time holographic interferometry using optical phase conjugation in photorefractive materials and direct spatial phase reconstruction, *Opt. Eng., 35*, 1383–1388 (1996).

Holographic Inteferometry

1. R. L. Powell and K. A. Stetson, Interferometric vibration analysis by wavefront reconstruction, *J. Opt. Soc. Am.*, *55*, 1593–1598 (1965).
2. K. A. Stetson and R. L. Powell, Interferometric Hologram evaluation and real-time vibration analysis of diffuse objects, *J. Opt. Soc. Am.*, *55*, 1694–1695 (1965).
3. R. J. Collier, E. T. Doherty and K. S. Pennington, Application of moiré techniques to holography, *Appl. Phys. Lett.*, *7*, 223-225 (1965).
4. R. E. Brooks, L. O. Heflinger and R. F. Wuerker, Interferometry with a holographically reconstructed comparison beam, *Appl. Phys. Lett.*, *7*, 248–249 (1965).
5. M. H. Horman, An application of wavefront reconstruction to interferometry, *Appl. Opt.*, *4*, 333–336 (1965).
6. L. H. Tanner, Some applications of holography in fluid mechanics, *J. Sci. Instrum, 43*, 81–83 (1966).
7. B. P. Hildebrand and K. A. Haines, Interferometric measurements using wavefront reconstruction technique, *Appl. Opt.*, *5*, 172–173 (1996).
8. K. A. Haines and B. P. Hildebrand, Surface-deformation measurement using the wavefront reconstruction technique, *Appl. Opt.*, *5*, 595–602 (1966).
9. L. O. Heflinger, R. F. Wuerker and R. E. Brooks, Holographic Interferometry, *J. Appl. Phys.*, *37*, 642–649 (1966).
10. J. M. Burch, A. E. Ennos and R. J. Wilton, Dual- and multiple-beam interferometry by wave front reconstruction, *Nature, 209*, 1015–1016 (1966).
11. B. P. Hildebrand and K. A. Haines, Multiple-wavelength and multiple-source holography applied for contour generation. *J. Opt. Soc. Am.*, *57*, 155–162 (1967).
12. E. Archbold, J. M. Burch and A. E. Ennos, The application of holography to the comparison of cylinder bores, *J. Sci. Instrum.*, *44*, 489–494 (1967).
13. L. H. Tanner, The scope and limitations of three-dimensional holography of phase objects, *J. Sci. Instrum.*, *44*, 1011–1014 (1967).
14. T. Tsuruta, N. Shiotake and Y. Itoh, Hologram interferometry using two reference beams, *Jpn. J. Appl. Phys.*, *7*, 1092–1100 (1967).

15. M. De and L. Sevigny, Three beam holographic interferometry, *Appl. Opt.*, *6*, 1665–1671 (1967).
16. E. B. Aleksandrov and A. M. Bonch-Bruevich, Investigation of surface strains by the hologram technique, *Sov. Phys. Tech, Phys.*, *12*, 258–265 (1967).
17. T. Tsuruta and N. Shiotake, Holographic generation of contour map of diffusely reflecting surface by using immersion method, *Jpn. J. Appl. Phys.*, *6*, 661–662 (1967).
18. J. S. Zelenka and J. R. Varner, A new method for generating depth contours holographically, *Appl. Opt.*, *7*, 2107–2110 (1968).
19. E. Archbold and A. E. Ennos, Observation of surface vibration modes by stroboscopic hologram interferometry, *Nature, 217*, 942-943 (1968).
20. P. Shajenko and C. D. Johnson, Stroboscopic holographic interferometry, *Appl. Phys.*, *13*, 44–46 (1968).
21. B. M. Watrasiewicz and P. Spicer, Vibration analysis by stroboscopic holography, *Nature, 217*, 1142–1143 (1968).
22. J. W. C. Gates, Holographic phase recording by interference between reconstructed wavefronts from separate holograms, *Nature, 220*, 473–474 (1968).
23. G. S. Ballard, Double exposure holographic interferometry with separate reference beams, *J. Appl. Phys.*, *39*, 4846–4848 (1968).
24. W. G. Gottenberg, Some applications of holographic interferometry, *Exp. Mech.*, *8*, 405–410 (1968).
25. M. A. Monahan and K. Bromley, Vibration analysis by holographic interferometry *J. Acous. Soc. Amer.*, *44*, 1225–1231 (1968).
26. A. E. Ennos, Measurement of in-plane surface strain by hologram interferometry, *J. Sci. Instrum.: Phys. E.*, *1*, 731–734 (1968).
27. O. Bryngdahl, Shearing interferometry by wavefront reconstruction, *J. Opt. Soc. Am, 58*, 865–871 (1968).
28. C. C. Aleksoff, Time average holography extended, *Appl. Phys. Letts.*, *14*, 23–24 (1969).
29. F. M. Motteir, Time average holography with triangular phase modulation of the reference wave, *Appl. Phys. Lett.*, *15*, 285–287 (1969).
30. R. M. Grant and G. M. Brown, Holographic non-destructive testing (HNDT), *Material Eval.*, *27*, 79–84 (1969).
31. G. M. Brown, R. M. Grant and G. W. Stroke, Theory of holographic interferometry, *J. Acoust. Soc. Am.*, *45*, 1166–1179 (1969).

32. K. Matsumoto, Holographic multiple beam interferometry, *J. Opt. Soc. Am.*, *59*, 777–778 (1969).
33. N.-E. Molin and K. A. Stetson, Measuring combination mode vibration patterns by hologram interferometry, *J. Sci. Instrum.; Phys. E, 2*, 609–612 (1969).
34. I. Yamaguchi and H. Saito, Application of holographic interferometry to the measurement of Poisson's ratio, *Jpn. J. Appl. Phys.*, *8*, 768–771 (1969).
35. W. Van Deelan and P. Nisenson, Mirror blank testing by real-time holographic interfermetry, *Appl. Opt.*, *8*, 951–955 (1969).
36. N. Abramson, The holodiagram: A practical device for making and evaluating holograms, *Appl. Opt.*, *8*, 1235–1240 (1969).
37. A. A. Friesem and C. M. Vest, Detection of micro fractures by holographic interferometry, *Appl. Opt.*, *8*, 1253–1254 (1969).
38. J. S. Zelenka and J. R. Varner, Multiple-index holographic contouring, *Appl. Opt.*, *8*, 1431–1434 (1969).
39. J. E. Sollid, Holographic interferometry applied to measurements of small static displacements of diffusely reflecting surfaces, *Appl. Opt.*, *8*, 1587–1595 (1969).
40. W. T. Welford, Fringe visibility and localization in hologram interferometry, *Opt. Commun.*, *1*, 123–125 (1969).
41. T. Tsuruta, N. Shiotake and Y. Itoh, Formation and localization of holographically produced interference fringes, *Opt. Acta, 16*, 723–733 (1969).
42. K. A. Stetson, A rigorous treatment of the fringes of hologram interferometry, *Optik, 29*, 386–400 (1969).
43. S. Mallick and M. L. Roblin, Shearing interferometry by wavefront reconstruction using single exposure, *Appl. Phys. Lett.*, *14*, 61–62 (1969).
44. S. Walles, On the concept of homologous rays in holographic interferometry of diffusely reflecting surfaces, *Opt. Acta, 17*, 899–913 (1970).
45. R. A. Jeffries, Two wavelength holographic interferometry of partially ionised plasmas, *Phys. Fluids, 13*, 210–212 (1970).
46. A. D. Wilson, Holographically observed torsion in a cylindrical shaft, Appl. Opt., *9*, 2093–2097 (1970).
47. R. C. Sampson, Holographic interferometry applications in experimental mechanics, *Exp. Mech.*, *10*, 313–320 (1970).
48. A. D. Wilson, Characteristic functions for time-average holography, *J. Opt. Soc. Am.*, *60*, 1068–1071 (1970).

49. K. A. Stetson. Effect of beam modulation on fringe loci and localization in time-average hologram interferometry, *J. Opt. Soc. Am.*, *60*, 1378–1384 (1970).
50. D. B. Neumann, C. F. Jacobson and G. M. Brown, Holographic technique for determining the phase of vibrating objects, *Appl. Opt.*, *9*, 1357–1362 (1970).
51. R. Pryputniewicz and K. A. Stetson, Determination of sensitivity vectors in hologram interferometry from two known rotations of the object, *Appl. Opt.*, *19*, 2201–2205 (1970).
52. K. S. Mustafin, V. A. Seleznev and E. I. Shtyrkov, Use of the nonlinear properties of a photoemulsion for enhancing the sensitivity of holographic interferometry, *Opt. Spectrosc.*, *28*, 638–640 (1970).
53. K. Matsumoto and M. Takashima, Phase difference amplification by non-linear holography, *J. Opt. Soc. Am.*, *60*, 30–33 (1970).
54. C. H. F. Velzel, Small phase differences in holographic interferometry, *Opt. Commun.*, *2*, 289–291 (1970).
55. A. D. Wilson, Computed time-average holographic interferometric fringes of a circular plate vibrating simultaneously in two rationally or irrationally related modes, *J. Opt. Soc. Am.*, *61*, 924–929 (1971).
56. F. Weigl, A generalized technique of two wavelength non diffuse holographic interferometry, *Appl. Opt.*, *10*, 187–192 (1971).
57. G. V. Ostrovskaya and Yu. I. Ostrovskii, Two-wavelength hologram method for studying the dispersion properties of phase objects, *Sov. Phys. Tech. Phys.*, *15*, 1890–1892 (1971).
58. R. D. Matulka and D. J. Collins, Determination of three-dimensional density fields from holographic interferograms, *J. Appl. Phys.*, *42*, 1109–1119 (1971).
59. L. A. Kersch, Advanced concepts of holographic nondestructive testing, *Mater. Eval.*, *29*, 125–129 (1971).
60. R. A. Ashton, D. Slovin and H. I. Gerritsen, Interferometric holography applied to elastic stress and surface corrosion, *Appl. Opt.*, *10*, 440–441 (1971).
61. A. D. Wilson, In-plane displacement of a stressed membrane with a hole measured by holographic interferometry, *Appl. Opt.*, *10*, 908–912 (1971).
62. C. C. Aleksoff, Temporally modulated holography, *Appl. Opt.*, *10*, 1329–1341 (1971).
63. J. P. Waters, Holographic inspection of solid propellant to liner bonds, *Appl. Opt.*, *10*, 2364–2365 (1971).

64. T. D. Dudderar and R. O'Regan, Measurement of the strain field near a crack tip in polymethylmethacrylate by holographic interferometry, *Exp. Mech.*, *11*, 49–56 (1971).
65. A. D. Wilson, C. H. Lee, H. R. Lominac and D. H. Strope, Holographic and analytic study of semi-clamped rectangular plate supported by struts, *Exp. Mech.*, *11*, 229–234 (1971).
66. W. Aung and R. O'Regan, Precise measurement of heat transfer using holographic interferometry, *Rev. Sci. Instrum.*, *42*, 1755–1759 (1971).
67. P. M. Boone, Determination of three orthogonal displacement components from one double-exposure hologram, *Opt. Laser Technol.*, *4*, 162–167 (1972).
68. C. H. Agren and K. A. Stetson, Measuring the resonances of treble-viol plates by hologram interferometry, *J. Acoust. Soc. Amer.*, *51*, 1971–1983 (1972).
69. P. Larinov, A. V. Lukin and K. S. Mustafin, Holographic inspection of shapes of unpolished surfaces, *Sov. J. Opt. Technol.*, *39*, 154–155 (1972).
70. B. U. Achia and D. W. Thompson, Real-time hologram-moiré interferometry for liquid flow visualization, *Appl. Opt.*, *11*, 953–954 (1972).
71. W. P. Chu, D. M. Robinson and J. H. Goad, Holographic nondestructive testing with impact excitation, *Appl. Opt.*, *11*, 1644–1645 (1972).
72. R. E. Rowlands and I. M. Daniel, Application of holography to anisotropic composite plates, *Exp. Mech.*, *12*, 75–82 (1972).
73. S. K. Dhir and J. P. Sikora, An improved method for obtaining the general-displacement field from a holographic interferogram, *Exp. Mech.*, *12*, 323–327 (1972).
74. T. R. Hsu and R. G. Moyer, Application of holography in high-temperature displacement measurements, *Exp. Mech.*, *12*, 431–432 (1972).
75. K. S. Mustafin and V. A. Seleznev, Holographic interferometry with variable sensitivity, *Opt. Spectrosc.*, *32*, 532–535 (1972).
76. P. Hariharan and Z. S. Hegedus, Simple multiplexing technique for double-exposure hologram interferometry, *Opt. Commun.*, *9*, 152–155 (1973).
77. D. W. Sweeny and C. M. Vest, Reconstruction of three-dimensional refractive index field from multidirectional interferometric data, *Appl. Opt.*, *12*, 2649–2664 (1973).

78. C. A. Sciammarella and J. A. Gilbert, Strain analysis of a disk subjected to diametral compression by means of holographic interferometry, *Appl. Opt.*, *12*, 1951–1956 (1973).
79. N. Abramson and H. Bjelkhagan, Industrial holographic measurements, *Appl. Opt.*, *11*, 2792–2796 (1973).
80. K. Grunwald, D. Kaletsch, V. Lehmann and H. Wachuta, Holographische Interferometrie und deren quantitative Auswertung, demonstriet am Beispiel zylindrischer GfK-Rohre, *Optik, 37*, 102–109 (1973).
81. A. D. Luxmoore, Holographic detection of cracks in concrete, *Nondestr. Test.*, *6*, 258–263 (1973).
82. R. Dandliker, B. Ineichen and F. M. Mottier, High resolution hologram interferometry by electronic phase measurement, *Opt. Commun.*, *9*, 412–416 (1973).
83. T. Matsumoto, K. Iwata and R. Nagata, Measuring accuracy of three dimensional displacements in holographic interferometry, *Appl. Opt.*, *12*, 961–967 (1973).
84. Y. Doi, T. Komatsu and T. Fujimoto, Shearing interferometry by holography, *Jpn. J. Appl. Phys.*, *12*, 1036–1042 (1973).
85. N. Abramson, Sandwich hologram interferometry: a new dimension in holographic comparison, *Appl. Opt.*, *13*, 2019–2025 (1974).
86. I. Prikryl, Localization of interference fringes in holographic interferometry, *Opt. Acta, 21*, 675–681 (1974).
87. M. Dubas and W. Schumann, Sur la détermination holographique de l'état de déformation á la surface d'un corps non-transparent, *Opt. Acta, 21*, 547–562 (1974).
88. K. A. Stetson, Fringe interpretation for hologram interferometry of rigid-body motions and homogeneous deformations, *J. Opt. Soc. Am.*, *64*, 1–10 (1974).
89. R. Jones and D. A. Bijl, Holographic interferometric study of the end effects associated with the four point bending technique for measuring Poisson's ratio, *J. Phys. E: J. Sci. Instrum.*, *7*, 357–358 (1974).
90. T. R. Hsu, Large-deformation measurements by real-time holographic interferometry, *Exp. Mech.*, *14*, 408–411 (1974).
91. J. P. Sikora and F. T. Mendenhall, Holographic vibration study of a rotating propeller blade, *Exp. Mech.*, *14*, 230–232 (1974).
92. E. R. Robertson and W. King, The technique of holographic interferometry applied to the study of transient stresses, *J. Strain Anal.*, *9*, 44–49 (1974).

93. A. R. Luxmoore, Holographic detection of cracks, *J. Strain Anal.*, *9*, 50–51 (1974).
94. T. Matsumoto, K. Iwata and R. Nagata, Measurement of deformation in a cylindrical shell by holographic interferometry, *Appl. Opt.*, *13*, 1080–1084 (1974).
95. V. Fossati Bellani and A. Sona, Measurement of three-dimensional displacements by scanning a double-exposure hologram, *Appl. Opt.*, *13*, 1337–1341 (1974).
96. S. Amadesi, F. Gori, R. Grella and G. Guattari, Holographic methods for painting diagnostics, *Appl. Opt.*, *13*, 2009–2013 (1974).
97. P. R. Wendendal and H. I. Bjelkhagen, Dynamics of human teeth in function by means of double-pulsed holography: an experimental investigation, *Appl. Opt.*, *13*, 2481–2485 (1974).
98. M. Dubas and W. Schumann, The determination of strain at a nontransparent body by holographic interferometry, *Opt. Acta, 21*, 547–562 (1974).
99. H. J. Raterink and R. L. van Renesse, Investigation of holographic interferometry, applied for the detection of cracks in large metal objects, *Optik, 40*, 193–200 (1974).
100. H. Spetzler, C. H. Scholz and C. P. J. Lu, Strain and creep measurements on rocks by holographic interferometry, *Pure Appl. Geophys.*, *112/3*, 571–582 (1974).
101. C. A. Sciammarella and T. Y. Chang, Holographic interferometry applied to the solution of shell problem, *Exp. Mech.*, *14*, 217–224 (1974).
102. E. Roth, Holographic interferometric investigations of plastic parts, in *Laser 75 Optoelectronics*, 168–174 (1975).
103. M. C. Collins and C. E. Watterson, Surface-strain measurements on a hemispherical shell using holographic interferometry, *Exp. Mech.*, *15*, 128–132 (1975).
104. K. Grunwald, W. Fritzsch and H. Wachuta, Quantitative holographische Verformungsmessungen an Kunststoff-und GFK Bauteilen, *Materialpruf*, *17*, 69–72 (1975).
105. Y. Chen, Holographic interferometry applied to rotating disk, *Trans. ASME, 42*, 499–512 (1975).
106. F. Able, A. L. Dancer, H. Fagot, R. B. Franke and P. Smigielski, Holographic interferometry applied to investigations of tympanic-membrane displacements in guinea-pig ears subjected to acoustical pulses, *J. Acoust. Soc. Am.*, *58*, 223–228 (1975).

107. J. D. Trolinger, Flow visualization holography, *Opt. Eng*, *14*, 470–481 (1975).
108. R. J. Radley Jr., Two-wavelength holography for measuring plasma electron density, *Phys. Fluids*, *18*, 175–179 (1975).
109. C. A. Sciammarella and J. A. Gilbert, Holographic interferometry applied to the measurement of displacements of interior points of transparent bodies, *Appl. Opt.*, *15*, 2176–2182 (1976).
110. C. G. H. Foster, Accurate measurement of Poisson's ratio in small samples, *Exp. Mech.*, *16*, 311–315 (1976).
111. H. D. Meyer and H. A. Spetzler, Material properties using holographic interferometry, *Exp. Mech.*, *16*, 434–438 (1976).
112. D. Dameron and C. M. Vest, Fringe sharpening and diffraction in nonlinear two-exposure holographic interferometry, *J. Opt. Soc. Am.*, *66*, 1418–1421 (1976).
113. N. Takai, M. Yamada and T. Idogawa, Holographic interferometry using a reference wave with a sinusoidally modulated amplitude, *Opt. Laser Technol.*, *8*, 21–23 (1976).
114. J. R. Crawford and R. Benson, Holographic interferometry of circuit board components, *Appl. Opt.*, *15*, 24–25 (1976).
115. P. Greguss, Holographic interferometry in biomedical sciences, *Opt. Laser Technol.*, *8*, 153–159 (1976).
116. N. Abramson, Holographic contouring by translation, *Appl. Opt.*, *15*, 1018–1022 (1976).
117. P. M. de Larminat and R. P. Wei, A fringe compensation technique for stress analysis by reflection holographic interferometry, *Exp. Mech.*, *16*, 241–248 (1976).
118. R. Dandliker, E. Marom and F. M. Mottier, Two-reference beam holographic interferometry, *J. Opt. Soc. Am.*, *66*, 23–30 (1976).
119. K. A. Stetson, Use of fringe vectors in hologram interferometry to determine fringe localization, *J. Opt. Soc. Am.*, *66*, 627–628 (1976).
120. P. Hariharan and Z. S. Hegedus, Two-hologram interferometry: a simplified sandwich technique, *Appl. Opt.*, *15*, 848–849 (1976).
121. G. E. Sommargran, Double-exposure holographic interferometry using common-path reference waves, *Appl. Opt.*, *16*, 1736–1741 (1977).
122. N. Abramson, Sandwich hologram interferometry: holographic studies of two milling machines, *Appl. Opt.*, *16*, 2521–2531 (1977).
123. D. C. Holloway, A. M. Patacca and W. L. Fourney, Application of holographic interferometry to a study of wave propagation in rock, *Exp. Mech.*, *17*, 281–289 (1977).

124. G. Wernicke and G. Frankowski, Untersuchung des Fliessbeginns einsatzgeharter und gaskarbonitrierter Stahl mit Hilfe der holographischen Interferometrie, *Die Technik*, *32*, 393–396 (1977).
125. C. M. Vest and D. W. Sweeney, Applications of holographic interferometry to non-destructive testing, *Int. Adv. Nondestru. Test.*, *5*, 17–21 (1977).
126. M. D. Mayer and T. E. Katyanagi, Holographic examination of a composite pressure vessel, *J. Test. Eval.*, *5*, 47–52 (1977).
127. S. Toyooka, Holographic interferometry with increased sensibility for diffusely reflecting objects, *Appl. Opt.*, *16*, 1054–1057 (1977).
128. H. Kohler, General formulation of the holographic interferometric evaluation methods, *Optik*, *47*, 469–475 (1977).
129. N. Abramson and H. Bjelkhagan, Pulsed sandwich holography, 2. Practical application, *Appl. Opt.*, *17*, 187–191 (1978).
130. D. Nobis and C. M. Vest, Statistical analysis of errors in holographic interferometry, *Appl. Opt.*, *17*, 2198–2204 (1978).
131. K. A. Stetson and I. R. Harrison, Computer-aided holographic vibration analysis for vectorial displacements of bladed disks, *Appl. Opt.*, *17*, 1733–1738 (1978).
132. P. M. De Larminat and R. P. Wei, Normal surface displacement around a circular hole by reflection holographic interferometry, *Exp. Mech.*, *18*, 74–80 (1978).
133. K. N. Petrov and Yu P. Presnyakov, Holographic interferometry of the corrosion process, *Opt. Spectrosc.*, *44*, 179–181 (1978).
134. J. M. Caussignac, Application of holographic interferometry to the study of structural deformations in Civil Engineering, *Proc. SPIE, 136*, 136–142 (1978).
135. A Felske, Holographic analysis of oscillations in squealing disk brakes, *Proc. SPIE, 136*, 148–155 (1978).
136. J. D. Dubourg, Application of holographic interferometry to testing of spun structures, *Proc. SPIE, 136*, 186–191 (1978).
137. P. Meyrueis, M. Pharok and J. Fontaine, Holographic interferometry in osteosynthesis, *Proc. SPIE, 136*, 202–205 (1978).
138. S. Toyooka, Computer mapping of the first and the second derivatives of plate deflection by using modulated diffraction gratings made by double-exposure holography, *Opt. Acta*, *25*, 991–1000 (1978).
139. K. A. Stetson, The use of an image-derotator in hologram interferometry and speckle photography of rotating objects, *Exp. Mech.*, *18*, 67–73 (1978).

140. K. D. Hinsch, Holographic interferometry of surface deformations of transparent fluids, *Appl. Opt.*, *17*, 3101–3108 (1978).

141. F. T. S. Yu, A. Tai and H. Chen, Multi-wavelength rainbow holographic interferometry, *Appl. Opt.*, *18*, 212–218 (1979).

142. Chandra Shakher and R. S. Sirohi, Fringe control techniques applied to holographic non-destructive testing, *Can. J. Phys.*, *57*, 2155–2160 (1979).

143. J. C. McBain, I. E. Horner and W. A. Stange, Vibration analysis of a spinning disk using image-derotated holographic interferometry, *Exp. Mech.*, *19*, 17–22 (1979).

144. G. von Bally, Otological investigations in living man using holographic interferometry, in *Holography in Medicine & Biology*, (ed. G. von Bally), Springer-Verlag, 18, 198–205 (1979).

145. S. Vikram and K. Vedam, Holographic interferometry of corrosion, *Optik*, *55*, 407–414 (1980).

146. S. Vikram, K. Vedam and E. G. Buss, Non-destructive evaluation of the strength of eggs by holography, *Poultry Sci.*, *59*, 2342–2347 (1980).

147. J. C. McBain, W. A. Stange and K. G. Harding, Real-time response of a rotating disk using image-derotated holographic interferometry, *Exp. Mech.*, *21*, 34–40 (1981).

148. W. F. Fagan, M.-A. Beeck and H. Kreitlow, The holographic vibration analysis of rotating objects using a reflective image derotator, *Opt. Lasers in Eng.*, *2*, 21–33 (1981).

149. C. A. Sciammarella, Holographic moiré, an optical tool for the determination of displacements, strains, contours, and slopes of surfaces, *Opt. Eng.*, *21*, 447–457 (1982).

150. M. Yonemura, Holographic contour generation by spatial frequency modulation, *Appl. Opt.*, *21*, 3652–3658 (1982).

151. H. J. Tiziani, Real-time metrology with BSO crystals, *Opt. Acta*, *29*, 463–470 (1982).

152. M. I. Marchant and M. B. Snell, Determination of flexural stiffness of thin plates from small deflection measurements using optical holography, *J. Strain Anal.*, *17*, 53–61 (1982).

153. A. E. Ennos and M. S. Virdee, Application of Reflection Holography to Deformation Measurement Problems, *Exp. Mech.*, *22*, 202–209 (1982).

154. P. Hariharan, B. F. Oreb and N. Brown, A digital phase-measurement system for real-time holographic interferometry, *Opt. Commun.*, *41*, 393–396 (1982).

155. T. Yatagai, S. Nakadate, M. Idesawa and H. Saito, Automatic fringe analysis using digital image processing techniques, *Opt. Eng.*, *21*, 432–435 (1982).
156. K. A. Stetson, Method of vibration measurements in heterodyne interferometry, *Opt. Lett.*, *7*, 233–234 (1982).
157. H. G. Leis, Vibration analysis of an 8-cylinder V-engine by time-average holographic interferometry, *SPIE*, *398*, 90–94 (1983).
158. R. Dandliker and R. Thalmann, Deformation of 3-D displacement and strain by holographic interferometry for non-plane objects, *Proc. SPIE*, *398*, 11–16 (1983).
159. G. Schonebeck, New holographic means to exactly determine coefficients of elasticity, *Proc. SPIE*, *398*, 130–136 (1983).
160. J. Geldmacher, H. Kreitlow, P. Steinlein and G. Sepold, Comparison of vibration mode measurements on rotating objects by different holographic methods, *Proc. SPIE*, *398*, 101–110 (1983).
161. S. Amadesi, A. D'Altorio and D. Paoletti, Single-two hologram interferometry: a combined method for dynamic tests on painted wooded statues, *J. Optics*, *14*, 243–246 (1983).
162. A. A. Antonov, Inspecting the level of residual stresses in welded joints by laser interferometry, *Weld Prod.*, *30*, 29–31 (1983).
163. J. L. Goldberg, A method of three-dimensional strain measurement on non-ideal objects using holographic interferometry, *Exp. Mech.*, *23*, 59–73 (1983).
164. H. P. Rossmanith, Determination of dynamic stress-intensity factors by holographic interferometry, *Opt. Lasers in Eng.*, *4*, 129–143 (1983).
165. P. Hariharan, B. F. Oreb and N. Brown, Real-time holographic interferometry: a microcomputer system for the measurement of vector displacement, *Appl. Opt.*, *22*, 876–880 (1983).
166. P. Lam, J. D. Gaskill and J. C. Wyant, Two-wavelength holographic interferometer, *Appl. Opt.*, *23*, 3079–3081 (1984).
167. Z. Füzessy and F. Gyimesi, Difference holographic interferometry: displacement measurement, *Opt. Eng.*, *23*, 780–783 (1984).
168. L. Wang, J. D. Hovanesian and Y. Y. Hung, A new fringe carrier method for the determination of displacement derivatives in hologram interferometry, *Opt. Lasers in Eng.*, *5*, 109–120 (1984).
169. R. J. Pryputniewicz, Heterodyne holography, applications in studies of small components, *Opt. Eng.*, *24*, 843–848 (1985)/*Opt. Eng.*, *24*, 849–854 (1985).

170. R. Thalmann and R. Dandliker, Heterodyne and quasi-heterodyne holographic interferometry, *Opt. Eng.*, *24*, 824–831 (1985).
171. P. Hariharan, Quasi-heterodyne hologram interferometry, *Opt. Eng.*, *24*, 632–638 (1985).
172. D. L. Mader, Holographic Interferometry on pipes: precision interpretation by least square fitting, *Appl. Opt.*, *24*, 3784–3790 (1985).
173. R. Pryputniewicz, Heterodyne holography, applications in study of small components, *Opt. Eng.*, *24*, 849–854 (1985).
174. H. Steinbichler, Holographic non-destructive testing with automatic evaluation, Automatic Optical Inspection (eds. L. R. Baker and H. J. Caulfield) *Proc. SPIE 645*, Cannes (1986).
175. W. Osten, Some consideration on the statistical error analysis in holographic interferometry with application to an optimized interferometry, *Opt. Acta*, *32*, 827–838 (1985).
176. A. Stimpfling and P. Smigielski, New method for compensating and measuring any motion of three-dimensional objects in holographic interferometry, *Opt. Eng.*, *24*, 821–823 (1985).
177. D. B. Neumann, Comparative holography: a technique for eliminating background fringes in holographic interferometry, *Opt. Eng.*, *24*, 625–627 (1985).
178. R. Pryputniewicz, Time average holography in vibration analysis, *Opt. Eng.*, *24*, 843-848 (1985).
179. Z. Fuzessy, F. Geimesi and J. Kornis, Comparison of two filament lamps by difference hologram interferometry, *Opt. Laser Technol.*, *18*, 318–320 (1986).
180. D. V. Nelson and J. McCrickerd, Residual-stress determination through combined use of holographic interferometry and blind-hole drilling, *Exp. Mech.*, *26*(4), 371–378 (1986).
181. S. Takemoto, Application of laser holographic techniques to investigate crustal deformations, *Nature*, *322*, 49–51 (1986).
182. T. Kreis, Digital holographic phase measurement using Fourier transform method, *J. Opt. Soc. Am.*, *A3*, 847–855 (1986).
183. D. Paoletti, G. Schirripa-Spagnolo and A. D'Altorio, Sandwich hologram for displacement derivatives, *Opt. Commun.*, *56*, 325–329 (1986).
184. D. W. Watt and C. M. Vest, Digital interferometry for flow visualization, *Exp. Fluids*, *5*, 401–406 (1987).

185. P. Del Vo and M. L. Rizzi, Vibrational testing of an X-ray concentrator by holographic interferometry, *Proc. SPIE*, *814*, 357–364 (1987).

186. H. Kreitlow, T. Kreis and W. Jüptner, Holographic interferometry with reference beams modulated by the object motion, *Appl. Opt.*, *26*, 4256–4262 (1987).

187. D. E. Cuche, Determination of Poisson's ratio by the holographic moiré technique, *Proc. SPIE*, *1026*, 165–170 (1988).

188. P. J. Bryanston-Cross and J. W. Gardner, Application of holographic interferometry to the vibrational analysis of the harpsichord, *Opt. Laser Technol.*, *20*, 199–204 (1988).

189. L. Wang and J. Ke, The measurement of residual stresses by sandwich holographic interferometry, *Opt. Lasers in Eng.*, *9*, 111–119 (1988).

190. H. Kasprzak, H. Podbielska and G. von Bally, Human tibia rigidly examined in bending and torsion loading by using double-exposure holographic interferometry, *Proc. SPIE*, *1026*, 196–201 (1988).

191. P. K. Rastogi, P. Jacquot and L. Pflug, Holographic interferometry applied at subfreezing temperatures: study of damage in concrete exposed to frost action, *Opt. Eng.*, *27*, 172–178 (1988).

192. T. E. Carlsson, G. Bjarnholt, N. Abramson and D. C. Holloway, Holographic interferometry applied to a model study of ground vibrations produced from blasting, *Opt. Eng.*, *27*, 923–927 (1988)

193. R. Dandliker, R. Thalmann and D. Prongue, Two-wavelength laser interferometry using super-heterodyne detection, *Opt. Lett.*, 13, 339–341 (1988).

194. G. Lai and T. Yatagai, Dual-reference holographic interferometry with a double pulsed laser, *Appl. Opt.*, *27*, 3855–3858 (1988).

195. S. Fu and J. Chen, Phase difference amplification of double-exposure holograms, *Opt. Commun.*, *67*, 417–420 (1988).

196. N. V. Kukhtarev and V. V. Muravev, Dynamic holographic interferometry in photorefractive crystals, *Opt. Spectros.*, *64*, 656–659 (1988).

197. M. Kujawinska and D. W. Robinson, Multichannel phase-stepped holographic interferometry, *Appl. Opt.*, 27, 312–320 (1988).

198. J. Gryzagoridis, Holographic non-destructive testing of composites, *Opt. Laser Technol.*, *21*, 113–116 (1989).

199. E. Simova and V. Sainov, Comparative holographic moiré interferometry for nondestructive testing: comparison with conventional holographic interferometry, *Opt. Eng.*, *28*, 261–266 (1989).
200. A. J. Decker, Holographic interferometry with an injection sealed Nd:Yag laser and two reference beams, *Appl. Opt.*, *29*, 2697–2700 (1990).
201. M. Takeda, Spatial-carrier fringe-pattern analysis and its application to precision interferometry and profilometry: an overview, *Industrial Metrology*, *1*, 79-99 (1990).
202. K. A. Stetson, Use of sensitivity vector variations to determine absolute displacements in double-exposure hologram interferometry, *Appl. Opt.*, *29*, 502–504 (1990).
203. V. Balalov, V. S. Pisarev, V. P. Shchepinov and V. V. Yakovlev, Holographic interference measurements of 3-D displacement fields and their use in stress determination, *Opt. Spectros., 68, 75–78 (1990).*
204. D. W. Watt, T. S. Gross and S. D. Hening, Three illumination beam phase-shifted holographic interferometry study of thermally induced displacements on a printed wiring board, *Appl. Opt.*, *30*, 1617–1623 (1991).
205. Z. Fuzessy, Application of double pulse holgraphy for the investigations of machines and systems, *Physical Research*, *15*, 75–107, Akademie Verlag (1991).
206. M. R. Sajan, T. S. Radha, B. S. Ramprasad and E. S. R. Gopal, Measurement of corrosion rate of aluminum in sodium hydroxide using holographic interferometry, *Opt. Lasers in Eng.*, *15*, 183–188 (1991).
207. P. K. Rastogi, M. Barillot and G. H. Kaufmann, Comparative phase-shifting holographic interferometry, *Appl. Opt.*, *30*, 722–728 (1991).
208. R. C. Troth and J. C. Dainty, Holographic interferometry using anisotropic self-diffraction in $Bi_{12}SiO_{20}$, *Opt. Lett.*, *16*, 53–55 (1991).
209. S. L. Sochava, R. C. Troth and S. I. Stepanov, Holographic interferometry using –1 order diffraction in photorefractive $Bi_{12}SiO_{20}$ and $Bi_{12}TiO_{20}$ crystals, *J. Opt. Soc. Am.*, *B9*, 1521–1527 (1992).
210. A. M. Lyalikov and A. F. Tuev, Multiwavelength single-exposure holographic interferometry, *Opt. Spectros.*, *73*, 227–229 (1992).

211. C. S. Vikram and W. K. Witherow, Critical needs of fringe order accuracies in two-color holographic interferometry, *Exp. Mech.*, *32*, 74–77 (1992).
212. M.-A. Beeck, Pulsed holographic vibration analysis on high speed rotating objects: fringe formation, recording techniques and practical applications, *Opt. Eng.*, *31*, 553–561 (1992).
213. W. Jüptner, J. Geldmacher, Th. Bischof and Th. Kreis, Measurement of the deformation of a pressure vessel above a weld point, *Proc. SPIE*, *1756*, 98–105 (1992).
214. G. Brown, J. W. Forbes, Mitchell M. Marchi and R. R. Wales, Hologram interferometry in automotive component vibration testing, *Proc. SPIE, 1756*, 146–152 (1992).
215. S. W. Biederman and R. P. Pryputniewicz, Holographic study of vibrations of a wing section, *Proc. SPIE*, *1756*, 153–163 (1992).
216. J. Woisetschlaeger, D. B. Sheffer, H. Mikati, K. Somasundaram, C. Wm. Loughry, S. K. Chawla and P. J. Wesolowski, Breast cancer detection by holographic interferometry, *Proc. SPIE*, *1756*, 176–183 (1992).
217. X. Wang, R. Magnusson and A. Haji-Sheikh, Real-time interferometry with photorefractive reference holograms, *Appl. Opt.*, *32*, 1983–1986 (1993).
218. P. Zanetta, G. P. Solomos, M. Zurn and A. C. Lucia, Holographic detection of defects in composites, *Opt. Laser Technol.*, *25*, 97–102 (1993).
219. M. M. Ratnam and W. T. Evans, Comparison of measurement of piston deformation using holographic interferometry and finite elements, *Exp. Mech.*, *12*, 336–343 (1993).
220. C. M. E. Holden, S. C. J. Parker and P. J. Bryanston-Cross, Quantitative three-dimensional holographic interferometry for flow field analysis, *Opt. Lasers in Eng.*, *19*, 285–298 (1993).
221. J. D. Trolinger and J. C. Hsu, Flowfield diagnostics by holographic interferometry and tomography, *Fringe '93*, (eds. W. Jüptner and W. Osten) 423–439 Akademie Verlag (1993).
222. A. Kreuttner, B. Lau, A. Mann, E. Mattes, R. Miller, M. Stuber, K. Stocker, Holographic interferometry for display of shock wave induced deformations and vibrations: a contribution to laser lithotripsy, *Lasers in Med. Sci.*, *8*, 211–220 (1993).
223. P. K. Rastogi, Holographic comparison between waves scattered from two physically distinct rough surfaces, *Opt. Eng.*, *33*, 3484–3485 (1994).

224. Q. Huang, J. A. Gilbert and H. J. Caulfield, Holographic interferometry using substrate guided waves, *Opt. Eng.*, *33*, 1132–1137 (1994).
225. U. Schnars and W. Jüptner, Digital recording and reconstruction of holograms in hologram interferometry and shearography, *Appl. Opt.*, *33*, 4373–4377 (1994).
226. U. Schnars, Direct phase determination in hologram interferometry with use of digitally recorded holograms, *J. Opt. Soc. Am.*, *A11*, 2011–2015 (1994).
227. S. V. Miridonov, A. A. Kamshilin and E. Barbosa, Recyclable holographic interferometer with a photorefractive crystal: optical scheme optimization, *J. Opt. Soc. Am.*, *A11*, 1780–1788 (1994).
228. D. Dirksen and G. von Bally, Holographic double-exposure interferometry in near real time with photorefractive crystals, *J. Opt. Soc. Am.*, *A11*, 1858–1863 (1994).
229. H. Kasprzak, W. N. Forester and G. von Bally, Holographic measurement of changes of the central corneal curvature due to interocular pressure differences, *Opt. Eng.*, *33*, 198–203 (1994).
230. N. Ninane and M. P. George, Holographic interferometry using two-wavelength holography: the measurement of large deformations, *Appl. Opt.*, *34*, 1923–1928 (1995).
231. Th. Kreis, R. Biederman and W. Jüptner, Evaluation of holographic interference patterns by artificial neural networks, *Proc. SPIE*, *2544*, 11–24 (1995).
232. B. Pouet and S. Krishnaswamy, Dynamic holographic interferometry by photorefractive crystals for quantitative deformation measurements, *Appl. Opt.*, *35*, 787–794 (1996).
233. V. I. Vlad, D. Malacara-Hernandez and A. Petris, Real-time holographic interferometry using optical phase conjugation in photorefractive materials and direct spatial phase modulation, *Opt. Eng.*, *35*, 1383–1387 (1996).
234. J. Zhu, Optical Nondestructive Examination of Glass-fiber Reinforced Plastic Honeycomb Structures, *Opt. Lasers in Eng.*, *25*, 133–143 (1996).
235. M. F. Ralea and N. N. Rosu, Laser holographic interferometry for investigations of cylindrical transparent tubes, *Opt. Eng.*, *35*, 1393–1395, (1996).

Speckle Metrology

1. J. A. Leendertz, Interferometric displacement measurement on scattering surface utilizing speckle effect, *J. Phys. E: Sci. Instrum.*, *3*, 214–218 (1970).
2. J. N. Butters and J. A. Leendertz, Speckle pattern and holographic techniques in engineering metrology, *Opt. Laser Technol.*, *3*, 26–30 (1971).
3. J. N. Butters and J. A. Leendertz, A double exposure technique for speckle pattern interferometry, *J. Phys.*, *4*, 272–279 (1971).
4. J. N. Butters and J. A. Leendertz, Holographic and video techniques applied to engineering measurement, *Trans. Inst. Meas. Control*, *4*, 349–354 (1971).
5. A. Macovski, S. D. Ramsey and L. F. Schaefer, Time-lapse interferometry and contouring using television systems, *Appl. Opt.*, *10*, 2722–2727 (1971).
6. H. J. Tiziani, Application of speckling for in-plane vibration analysis, *Opt. Acta*, *18*, 891–902 (1971).
7. H. J. Tiziani, A study of the use of laser speckle to measure small tilts of optically rough surfaces accurately, *Opt. Commun.*, *5*, 271–276 (1972).
8. D. E. Duffy, Moiré gauging of in-plane displacement using double aperture imaging, *Appl. Opt.*, *11*, 1778–1781 (1972).
9. E. Archbold and A. E. Ennos, Displacement measurement from double-exposure laser photographs, *Opt. Acta*, *19*, 253–273 (1972).
10. J. A. Leendertz and J. N. Butters, An image-shearing speckle-pattern interferometer for measuring bending moments, *J. Phys. E: Sci. Instrum.*, *6*, 1107–1110 (1973).
11. Y. Y. Hung, A speckle-shearing interferometer: A tool for measuring derivatives of surface displacements, *Opt. Commun.*, *11*, 132–135 (1974).
12. H. M. Pedersen, O. J. Lokberg and B. M. Foerre, Holographic vibration measurement using a TV speckle interferometer with silicon target vidicon, *Opt. Commun.*, *12*, 421–426 (1974).
13. K. A. Stetson, Analysis of double-exposure speckle photography with two-beam illumination, *J. Opt. Soc. Amer.*, *64*, 857–861 (1974).
14. G. Cloud, Practical speckle interferometry for measuring in-plane deformation, *Appl. Opt.*, *14*, 878–884 (1975).

15. P. M. Boone, Determination of slope and strain contours by double-exposure shearing interferometry, *Exp. Mech.*, *15*, 295–302 (1975).
16. M. Francon, P. Koulev and M. May, Detection of small displacements of diffuse objects illuminated by a speckle pattern, *Opt. Commun.*, *13*, 138–141 (1975).
17. R. Jones and J. N. Butters, Some observations on the direct comparison of the geometry of two objects using speckle pattern interferometric contouring, *J. Phys. E: Sci. Instrum.*, *8*, 231–234 (1975).
18. C. Forno, White-light speckle photography for measuring deformation, strain, and shape, *Opt. Laser Technol.*, *7*, 217–221 (1975).
19. D. A. Gregory, Basic physical principles of defocused speckle photography: a tilt topology inspection technique, *Opt. Laser Technol.*, *8*, 201–213 (1976).
20. R. Jones, The design and application of a speckle pattern interferometer for total plane strain field measurement, *Opt. Laser Technol.*, *8*, 215–219 (1976).
21. R. P. Khetan and F. P. Chiang, Strain analysis by one-beam laser speckle interferometry. 1: Single aperture method, *Appl. Opt.*, *15*, 2205–2215 (1976).
22. F. P. Chiang and R. M. Juang, Vibration analysis of plate and shell by laser speckle interferometry, *Opt. Acta*, *23*, 997–1009 (1976).
23. K. Hogmoen and O. J. Lokberg, Detection and measurement of small vibrations using electronic speckle pattern interferometry, *Appl. Opt.*, *16*, 1869–1875 (1977).
24. K. P. Miyake and H. Tamura, Observation and explanation of the localization of fringes formed with double-exposure speckle photograph, in *Applications of Holography and Optical Data Processing*, (eds. E. Marrom, A. A. Friesem and E. Wiener), Pergamon Press, London, pp. 333–339 (1977).
25. C. Wykes, De-correlation effects in speckle-pattern interferometry, *Opt. Acta*, *24*, 517–532 (1977).
26. T. J. Cookson, J. N. Butters and H. C. Pollard, Pulsed lasers in electronic speckle pattern interferometry, *Opt. Laser Technol.*, *10*, 119–124 (1978).
27. J. Brdicko, M. D. Olson and C. R. Hazell, Theory for surface displacement and strain measurements by laser speckle interferometry, *Opt. Acta*, *25*, 963–989 (1978).

28. Y. Y. Hung, I. M. Daniel and R. E. Rowlands, Full-field optical strain measurement having postrecording sensitivity and direction selectivity, *Exp Mech.*, *18*, 56–60 (1978).
29. O. J. Lokberg, K. Hogmoen and O. M. Holje, Vibration measurement on the human ear drum in vivo, *Appl. Opt.*, *18*, 763–765 (1979).
30. J. Brdicko, M. D. Olson and C. R. Hazell, New aspects of surface displacement and strain analysis by speckle interferometry, *Exp. Mech.*, *19*, 160–165 (1979).
31. Y. Y. Hung and A. J. Durelli, Simultaneous measurement of three displacement derivatives using a multiple image-shearing interferometric camera, *J. Strain Anal.*, *14*, 81–88 (1979).
32. J. Holoubek and J. Ruzek, A strain analysis technique by scattered-light speckle photography *Opt. Acta*, *26*, 43–54 (1979).
33. G. E. Slettemoen, General analysis of fringe contrast in electronic speckle pattern interferometry, *Opt. Acta*, *26*, 313–327 (1979).
34. R. P. Khetan and F. P. Chiang, Strain analysis by one-beam laser speckle interferometry. 2: Multiaperture method, *Appl. Opt.*, *18*, 2175–2186 (1979).
35. O. J. Lokberg, Electronic speckle pattern interferometry, *Phys. Technol.*, *11*, 16–22 (1980).
36. G. H. Kaufmann, On the numerical processing of speckle photograph fringes, *Opt. Laser Technol.*, *12*, 207–209 (1980).
37. G. E. Slettemoen, Electronic speckle pattern interferometric system based on a speckle reference beam, *Appl. Opt.*, *19*, 616–623 (1980).
38. S. Nakadate, T. Yatagai and H. Saito, Electronic speckle-pattern interferometry using digital image processing techniques, *Appl. Opt.*, *19*, 1879–1883 (1980).
39. S. Nakadate, T. Yatagai and H. Saito, Digital speckle-pattern shearing interferometry, *Appl. Opt.*, *19*, 4241–4246 (1980).
40. G. H. Kaufmann, A. E. Ennos, B. Gale and D. J. Pugh, An electro-optical read-out system for analysis of speckle photographs, *J. Phys. E: Sci. Instrum.*, *13*, 579–584 (1980).
41. I. Yamaguchi, Speckle displacement and decorrelation in the diffraction and image fields for small object deformation, *Opt. Acta*, *28*, 1359–1376 (1981).
42. O. J. Lokberg and G. E. Slettemoen, Interferometric comparison of displacements by electronic speckle pattern interferometry, *Appl. Opt.*, *20*, 2630–2634 (1981).

43. I. Yamaguchi, A laser-speckle strain gauge, *J. Phys. E: Sci. Instrum.*, *14*, 1270–1273 (1981).
44. G. K. Jaisingh and F. P. Chiang, Contouring by laser speckle, *Appl. Opt.*, *20*, 3385–3387 (1981).
45. C. S. Vikram and K. Vedam, Speckle photography of lateral sinusoidal vibrations: error due to varying halo intensity, *Appl. Opt.*, *20*, 3388–3391 (1981).
46. R. Jones and C. Wykes, General parameters for the design and optimization of electronic speckle pattern interferometers, *Opt. Acta.*, *28*, 949–972 (1982).
47. C. Wykes, Use of electronic speckle pattern interferometry (ESPI) in the measurement of static and dynamic surface displacements, *Opt. Eng.*, *21*, 400–406 (1982).
48. R. Krishna Murthy, R. S. Sirohi and M. P. Kothiyal, Detection of defects in plates and diaphragms using split lens speckle-shearing interferometer, *NDT International (UK)*, *15*, 329–33 (1982).
49. A. Asundi and F. P. Chiang, Theory and applications of the white light speckle method for strain analysis, *Opt. Eng.*, 21, 570–580 (1982).
50. R. K. Mohanty, C. Joenathan and R. S. Sirohi, Speckle shear interferometry with double dove prisms, *Opt. Commun.*, *47*, 27–30 (1983).
51. T. D. Dudderar, J. A. Gilbert, A. J. Boehnlein and M. E. Schultz, Application of fiber optics to speckle metrology—a feasibility study, *Exp. Mech.*, *23*, 289–297 (1983).
52. D. K. Sharma, R. S. Sirohi and M. P. Kothiyal, Simultaneous measurement of slope and curvature with three aperture speckle interferometer, *Appl. Opt.*, *23*, 1542–46 (1984).
53. D. W. Robinson, Automatic fringe analysis with a computer image-processing system, *Appl. Opt.*, *22*, 2169–2176 (1983).
54. Chandra Shakher and G. Venkata Rao, Use of holographic optical elements in speckle metrology, *Appl. Opt.*, *23*, 4592–4595 (1984).
55. R. Krishna Murthy, R. K. Mohanty, R. S. Sirohi and M. P. Kothiyal, Radial speckle shearing interferometer and its engineering applications, *Optik*, *67*, 85–94 (1984).
56. C. Joenathan, R. K. Mohanty and R. S. Sirohi, On the methods of multiplexing in speckle shear interferometry, *Optik*, *69*, 8–12 (1984).

57. R. K. Mohanty, C. Joenathan and R. S. Sirohi, Speckle interferometric methods of measuring out of plane displacements, *Opt. Letts.*, *9*, 475–477 (1984).
58. R. S. Sirohi, Speckle shear interferometry: a review, *J. Opt. (India)*, *13*, 95–113 (1984).
59. B. P. Holownia, Non-destructive testing of overlap shear joints using electronic speckle pattern interferometry, *Opt. Lasers in Eng.*, *6*, 79–90 (1985).
60. C. Joenathan, R. K. Mohanty, and R. S. Sirohi, Hololens in speckle and speckle shear interferometry, *Appl. Opt.*, *24*, 1294–1298 (1985).
61. R. K. Mohanty, C. Joenathan and R. S. Sirohi, Multiplexing in speckle shear interferometry: an optimal method, *Appl. Opt.*, *24*, 2043–44 (1985).
62. S. Nakadate and H. Saito, Fringe scanning speckle-pattern interferometry, *Appl. Opt.*, *24*, 2172–2180 (1985).
63. K. Creath, Phase-shifting speckle interferometry, *Appl. Opt.*, *24*, 3053–3058 (1985).
64. R. K. Mohanty, C. Joenathan and R. S. Sirohi, Speckle and speckle shear interferometers combined for simultaneous determination of out of plane displacement and slope, *Appl. Opt.*, *24*, 3106–3109 (1985).
65. O. J. Lokberg, J. T. Malmo and G. E. Slettemoen, Interferometric measurements of high temperature objects by electronic speckle pattern interferometry, *Appl. Opt.*, *24*, 3167–3172 (1985).
66. K. A. Stetson and W. R. Brohinsky, Electrooptic holography and its application to hologram interferometry, *Appl. Opt.*, *24*, 3631–3637 (1985).
67. R. Erbeck, Fast image processing with a microcomputer applied to speckle photography, *Appl. Opt.*, *24*, 3838–3841 (1985).
68. F. P. Chiang and D. W. Li, Random (speckle) patterns for displacement and strain measurement: some recent advances, *Opt. Eng.*, *24*, 936–943 (1985).
69. K. Creath and G. E. Slettemoen, Vibration-observation techniques for digital speckle-pattern interferometry, *J. Opt. Soc. Amer. A*, *2*, 1629–1636 (1985).
70. C. Joenathan, R. K. Mohanty and R. S. Sirohi, Contouring by speckle interferometry, *Opt. Letts.*, *10*, 579–81 (1985).
71. R. K. Mohanty, C. Joenathan and R. S. Sirohi, NDT speckle rotational shear interferometry, *NDT Int.* (*UK*), *18*, 203–05 (1985).

72. R. K. Mohanty, C. Joenathan and R. S. Sirohi, High sensitivity tilt measurement by speckle shear interferometry, *Appl. Opt.*, *25*, 1661–64 (1986).
73. J. R. Tyrer, Critical review of recent developments in electronic speckle pattern interferometry, *Proc. SPIE*, *604*, 95–111 (1986).
74. J. M. Huntley, An image processing system for the analysis of speckle photographs, *J. Phys. E: Sci. Instrum.*, *19*, 43–49 (1986).
75. C. Joenathan, C. S. Narayanamurthy and R. S. Sirohi, Radial and rotational slope contours in speckle shear interferometry, *Opt. Commun.*, *56*, 309–12 (1986).
76. S. Nakadate, Vibration measurement using phase-shifting speckle pattern interferometry, *Appl. Opt.*, *25*, 4162–4167 (1986).
77. C. Joenathan and R. S. Sirohi, Elimination of errors in speckle photography, *Appl. Opt.*, *25*, 1791–1794 (1986).
78. A. R. Ganesan, C. Joenathan and R. S. Sirohi, Real-time comparative digital speckle pattern interferometry, *Opt. Commun.*, *64*, 501–506 (1987).
79. F. Ansari and G. Ciurpita, Automated fringe measurement in speckle photography, *Appl. Opt.*, *26*, 1688–1692 (1987).
80. R. Hoefling and W. Osten, Displacement measurement by image-processed speckle patterns, *J. Mod. Opt.*, 34, 607–617 (1987).
81. O. J. Lokberg and J. T. Malmo, Detection of defects in composite materials by TV holography, *NDT Int. (UK)*, *21*, 223–228 (1988).
82. S. Winther, 3D Strain Measurements using ESPI, *Opt. Lasers Eng.*, *8*, 45–57 (1988).
83. M. Owner-Petersen and P. Damgaard Jensen, Computer-aided electronic speckle pattern interferometry (ESPI): deformation analysis by fringe manipulation, *NDT Int.*, *21*, 422–426 (1988).
84. C. Joenathan, C. S. Narayanamurthy and R. S. Sirohi, Localization of fringes in speckle photography that are due to axial motion of the diffuse object, *J. Opt. Soc. Am. (A)*, *5*, 1035–1040 (1988).
85. A. R. Ganesan and R. S. Sirohi, New method of contouring using digital speckle pattern interferometry (DSPI), *Proc. SPIE*, *954*, 327–332 (1988).
86. R. W. T. Preater, Measuring rotating component in-plane strain using conventional pulsed ESPI and optical fibres, *Proc. SPIE*, *952*, 245–250 (1988).
87. O. J. Lokberg and J. T. Malmo, Long-distance electronic speckle pattern interferometry, *Opt. Eng.*, *27*, 150–156 (1988).

88. Boxiang Lu, H. Abendroth, H. Eggers, E. Ziolkowski and X. Yang, Real time investigation of rotating objects using ESPI system, *Proc. SPIE, 1026*, 218–221 (1988).
89. V. Parthiban, C. Joenathan and R. S. Sirohi, A simple inverting interferometer with holo elements, *Appl. Opt., 27*, 1913–1914 (1988).
90. A. R. Ganesan, D. K. Sharma and M. P. Kothiyal, Universal digital speckle shearing interferometer, *Appl. Opt., 27*, 4731–4734 (1988).
91. R. J. Pryputniewicz and K. A. Stetson, Measurement of vibration patterns using electro-optic holography, *Proc. SPIE, 1162*, 456–467 (1989).
92. J. Davies and C. Buckberry, Applications of fibre optic TV holography system to the study of large automotive structures, *Proc. SPIE, 1162*, 279–291 (1989).
93. V. M. Chaudhari, A. R. Ganesan, Chandra Shakher, P. B. Godbole and R. S. Sirohi, Investigations of in-plane stresses on bolted flange joint using digital speckle pattern interferometry, *Opt. Lasers in Eng., 11*, 257–264 (1989).
94. E. Vikhagen, Vibration measurement using phase shifting TV-holography and digital image processing, *Opt. Commun., 69*, 214–218 (1989).
95. M. Kujawińska, A. Spik, and D. W. Robinson, Quantitative analysis of transient events by ESPI, *Proc, SPIE, 1121*, 416–423 (1989).
96. W. Arnold and K. D. Hinsch, Parallel optical evaluation of double-exposure records in optical metrology, *Appl. Opt., 28*, 726–729 (1989).
97. J. M. Huntley, Speckle photography fringe analysis: assessment of current algorithms, *Appl. Opt., 28*, 4316–4322 (1989).
98. K. D. Hinsch, Fringe positions in double-exposure speckle photography, *Appl. Opt., 28*, 5298–5304 (1989).
99. A. J. Moore and J. R. Tyrer, An electronic speckle pattern interferometer for complete in-plane displacement measurement, *Meas. Sci. Technol., 1*, 1024–1030 (1990).
100. R. P. Tatam, J. C. Davies, C. H. Buckberry and J. D. C. Jones, Holographic surface contouring using wavelength modulation of laser diodes, *Opt. Laser Technol., 22*, 317–321 (1990).

101. G. Gulker, K. Hinsch, C. Holscher, A. Kramer and H. Neunaber, Electronic speckle pattern interferometry system for in situ deformation monitoring on buildings, *Opt. Eng.*, *29*, 816–820 (1990).
102. E. Vikhagen, Nondestructive testing by use of TV holography and deformation phase gradient calculation, *Appl. Opt.*, *29*, 137–144 (1990).
103. M. Owner-Petersen, Decorrelation and fringe visibility: on the limiting behavior of various electronic speckle-pattern correlation interferometers, *J. Opt. Soc. Amer. A*, *8*, 1082–1089 (1991).
104. M. Owner-Petersen, Digital speckle pattern shearing interferometry: limitations and prospects, *Appl. Opt.*, *30*, 2730–2738 (1991).
105. C. Joenathan and B. M. Khorana, A simple and modified ESPI system, *Optik, 88*, 169–171 (1991).
106. H. Kadono, S. Toyooka and Y. Iwasaki, Speckle-shearing interferometry using a liquid-crystal cell as a phase modulator, *J. Opt. Soc. Amer. A*, *8*, 2001–2008 (1991).
107. R. Hoefling, P. Aswendt, W. Totzauer and W. Jüptner, DSPI: A tool for analysing thermal strains on ceramic and composite materials, *Proc. SPIE*, *1508*, 135–142 (1991).
108. S. Takemoto, Geophysical applications of holographic and ESPI techniques, *Proc. SPIE*, *1553*, 168–174 (1991).
109. M. M. Ratnam, W. T. Evans and J. R. Tyrer, Measurement of thermal expansion of a piston using holographic and electronic speckle pattern interferometry, *Opt. Eng.*, *31*, 61–69 (1992).
110. S. Ellingsrud and G. O. Rosvold, Analysis of a data-based TV-holography system used to measure small vibration amplitudes, *J. Opt. Soc. Amer. A*, *9*, 237–251 (1992).
111. R. R. Vera, D. Kerr and F. M. Santoyo, Electronic speckle contouring, *J. Opt. Soc. Amer. A*, *9*, 2000–2008 (1992).
112. R. S. Sirohi and N. Krishna Mohan, Speckle interferometry for deformation measurement, *J. Mod. Opt.*, *39*, 1293–1300 (1992).
113. A. R. Ganesan, B. C. Tan and R. S. Sirohi, Tilt measurement using digital speckle pattern interferometry, *Opt. Laser Technol.*, *24*, 257–261 (1992).
114. G. Gulker, O. Haack, K. D. Hinsch, C. Holscher, J. Kuls and W. Platen, Two-wavelength electronic speckle-pattern interferometry for the analysis of discontinuous deformation fields, *Appl. Opt*, *31*, 4519–4521 (1992).
115. R. Spooren, Double-pulse subtraction TV holography, *Opt. Eng.*, *31*, 1000–1007 (1992).

116. R. Rodriguez-Vera, D. Kerr and F. Mendoza-Santoyo, Electronic speckle contouring, *J. Opt. Soc. Amer. A*, *9*, 2000–2007 (1992).
117. X. Peng, H. Y. Diao, Y. L. Zou and H. Tiziani, A novel approach to determine decorrelation effect in a dual-beam electronic speckle pattern interferometer, *Optik*, *90*, 129–133 (1992).
118. I. Yamaguchi, Z. B. Liu, J. Kato and J. Y. Liu, Active phase-shifting speckle interferometer, in *Fringe '93*, (eds. W. Jüptner and W. Osten) pp. 103–108, Akademie Verlag (1993).
119. P. Aswendt and R. Hoefling, Speckle interferometry for analysing anisotropic thermal expansion—application to specimens and components, *Composites*, *24*, 611–617 (1993).
120. R. Spooren, A. Aksnes Dyrseth and M. Vaz, Electronic shear interferometry: application of a (double-) pulsed laser, *Appl. Opt.*, *32*, 4719–4727 (1993).
121. A. J. P. van Haasteren and H. J. Frankena, Real time displacement measurement using a multi-camera phase stepping interferometer, in *Fringe '93*, (eds. W. Jüptner and W. Osten) *19*, 417–422, Akademie Verlag, Bremen (1993).
122. G. Pedrini, B. Pfister and H. Tiziani, Double pulse-electronic speckle pattern interferometry, *J. Mod. Opt.*, *40*, 89–96 (1993).
123. N. Krishna Mohan, H. Saldner, N.-E. Molin, Electronic speckle pattern interferometry for simultaneous measurement of out-of-plane displacement and slope, *Opt. Letts.*, *18*, 1861–1863 (1993).
124. D. Paoletti and G. Schirripa Spagnolo, Automatic digital speckle pattern interferometry contouring in artwork surface inspection, *Opt. Eng.*, *32*, 1348–1353 (1993).
125. B. F. Pouet and S. Krishnaswamy, Additive/subtractive decorrelated electronic speckle pattern interferometry, *Opt. Eng.*, *32*, 1360–1368 (1993).
126. J. Kato, I. Yamaguchi and Q. Ping, Automatic deformation analysis by a TV speckle interferometer using a laser diode, *Appl. Opt.*, *32*, 77–83 (1993).
127. R. S. Sirohi and N. Krishna Mohan, In-plane displacement measurement configuration with two fold sensitivity, *Appl. Opt.*, *32*, 6387–90 (1993).
128. C. Joenathan, B. Franze and H. J. Tiziani, Oblique incidence and observation electronic speckle-pattern interferometry, *Appl. Opt.*, 33, 7307–7311 (1994).

129. A. R. Ganesan, P. Meinlschmidt and K. Hinsch, Vibration mode separation using comparative electronic speckle pattern interferometry (ESPI), *Opt. Commun.*, *107*, 28–34 (1994).
130. J. Sivaganthan, A. R. Ganesan, B. C. Tan, K. S. Law and R. S. Sirohi, Study of steel weldments using electronic speckle pattern interferometry, *J. Testing & Eval.*, *22*, 42–44 (1994).
131. P. R. Sreedhar, N. Krishna Mohan and R. S. Sirohi, Real time speckle photography with two wave mixing in photorefractive $BaTiO_3$ crystal, *Opt. Eng.*, *33*, 1989–95 (1994).
132. G. Jin, N. Bao and P. S. Chung, Applications of a novel phase shift method using a computer-controlled polarization mechanism, *Opt. Eng.*, *33*, 2733–2737 (1994).
133. O. J. Lokberg, Sound in flight: measurement of sound fields by use of TV holography, *Appl. Opt.*, *33*, 2574–2584 (1994).
134. H. M. Pedersen, O. J. Lokberg, H. Valx and G. Wang, Detection of non-sinusoidal periodic vibrations using phase modulated TV-holography, *Opt. Commun.*, *104*, 271–276 (1994).
135. R. S. Sirohi and N. Krishna Mohan, Role of lens aperturing in speckle metrology, *J. Sci. Ind. Res.*, *54*, 67–74 (1995).
136. G. Gulker and K. D. Hinsch, Detection of microstructure changes by electronic speckle pattern interferometry, *Opt. Lasers in Eng.*, *26*, 1348–1353 (1996).
137. R. S. Sirohi, N. Krishna Mohan and T. Santhanakrishnan, Optical configuration for measurement in speckle interferometry, *Opt. Letts.*, *21*, 1958–1959 (1996).
138. N. Krishna Mohan, T. Santhanakrishnan, P. Senthilkumaran, and R. S. Sirohi, Simultaneous implementation of Leendertz and Duffy methods for in-plane displacement measurement, *Opt. Commun.*, *124*, 235–239 (1996).
139. R. S. Sirohi and N. Krishna Mohan, An in-plane insensitive multi-aperture speckle shear interferometer for slope measurement, *Opt. Laser Technol.*, *29*, 415–417 (1997).
140. R. S. Sirohi, J. Burke, H. Helmers and K. D. Hinsch, Spatial phase-shifting for pure in-plane displacement and displacement derivatives measurement in electronic speckle pattern interferometry (ESPI), *Appl. Opt.*, *36*, 5787–5791 (1997).
141. P. K. Rastogi, Speckle shearing photography: a tool for direct measurement of surface strains, *Appl. Opt.*, *37*, 1292–1298 (1998).

Photo-elasticity

1. Albrecht Kuske and George Robertson, *Photoelastic Stress Analysis*, John Wiley and Sons, New York (1974).
2. J. W. Dally and W. F. Riley, *Experimental Stress Analysis*, McGraw-Hill, New York (1978).
3. H. Aben, *Integrated Photoelasticity*, McGraw-Hill, New York (1979).
4. A. Lagarde (ed.), *Optical Methods in Mechanics of Solids*, Sijthoff en Noordhoff, Alphen aan den Rijn, The Netherlands (1981).
5. L. S. Srinath, *Scattered Light Photo-elasticity*, Tata McGraw-Hill, New Delhi (1983).
6. Rajpal S. Sirohi and Mahendra P. Kothiyal, *Optical Components, Systems, and Measurement Techniques*, Marcel Dekker, Inc., New York (1991).
7. A. S. Kobayashi (ed.), *Handbook on Experimental Mechanics*, Society for Experimental Mechanics, (1993).
8. D. C. Drucker, The method of oblique incidence in photoelasticity, *Proc. SESA, VIII*, *1*, 51–66 (1950).
9. K. Ito, New model material for photoelasticity and photoplasticity, *Exp. Mech.*, *2*, 373–376 (1962).
10. A. S. Redner, New oblique-incidence method for direct photoelastic measurements of principal strains, *Exp. Mech.*, *3*(3), 67–72 (1963).
11. A. Robert and E. Guillemet, New scattered light method in three dimensional photoelasticity, *Brit. J. Appl. Physics*, *15*, 567–578 (1964).
12. M. Nishida and H. Saito, A new interferometric method of two dimensional stress analysis, *Proc. SESA*, *21* (2), 366–376 (1964).
13. Richard O'Regan, New method of determining strain on the surface of a body with photoelastic coating, *Exp. Mech.*, *5*, 241–246 (1965).
14. J. W. Dally and E. R. Erisman, An analytic separation method for photoelasticity, *Exp. Mech.*, *6*, 494–499 (1966).
15. A. J. Robert, New methods in photoelasticity, *Exp. Mech.*, *7*, 224–232 (1967).
16. M. E. Fourney, Application of holography to photoelasticity, *Exp. Mech.*, *8*, 33–38 (1968).
17. D. Hovanesian, V. Broic and R. L. Powell, A new experimental stress-optic method: stress-holo-interferometry, *Exp. Mech.*, *8*, 362–368 (1968).
18. H. H. H. Chau, Holographic interferometer for isopachic stress analysis, *Rev. Sci. Instrum.*, *39*, 1789–1792 (1968).

19. W. Wetzels, Holographie als hilfsmittel zur isopachenbestimmung, *Optik*, *27*, 271–272 (1968).
20. von Eric Hopp and G. Wutzke, The application of holography in plane photoelasticity, *Materialpruf.*, *11*, 409–415 (1969).
21. L. S. Srinath, Analysis of scattered-light methods in photoelasticity, *Exp. Mech.*, *9*, 463–468 (1969).
22. von Eric Hopp and G. Wutzke, Holographic determination of the principal stresses in plane models, *Materialpruf.*, *12*, 13–22 (1970).
23. R. C. Sampson, A stress-optic law for photoelastic analysis of orthotropic composites, *Exp. Mech.*, *10*, 210–215 (1970).
24. D. Post, Photoelastic fringe multiplication for ten fold increase in sensitivity, *Exp. Mech.*, *10*, 305–312 (1970).
25. D. C. Holloway and R. H. Johnson, Advancements in holographic photoelasticity, *Exp. Mech.*, *11*, 57–63 (1971).
26. B. Chatelain, Holographic photo-elasticity: independent observation of the isochromatic and isopachic fringes for a single model subjected to only one process, *Opt. Laser Technol.*, *5*, 201–204 (1973).
27. S. Redner, New automatic polariscope system, *Exp. Mech.*, *14*, 486–491 (1974).
28. J. Ebbeni, J. Coenen and H. Hermanne, New analysis of holophotoelastic patterns and their application, *J. Strain Anal.*, *11*, 11–17 (1976).
29. J. S. Parks and R. J. Sanford, On the role of material and optical properties in complete photoelastic analysis, *Exp. Mech.*, *16*, 441–447 (1976).
30. M. Arcan, Z. Hashin and A. Voloshin, A method to produce uniform plane stress states with applications, *Exp. Mech.*, *18*, 141–145 (1978).
31. J. F. Doyle and H. T. Denyluk, Integrated photoelasticity for axisymmetric problems, *Exp. Mech.*, *18*, 215–220 (1978).
32. J. W. Dally and R. J. Sanford, Classification of stress intensity Factors from isochromatic fringe patterns, *J. Exp. Mech.*, *18*, 441–448 (1978).
33. P. S. Theocaris and E. E. Goloutos, A unified interpretation of interferometric and holographic fringe patterns in photoelasticity, *J. Strain Anal.*, *13*, 95–102 (1978).
34. D. Post, Photoelasticity, *Exp. Mech.*, *19*, 176–192 (1979).

35. S. B. Mazurikiewicz and J. T. Pindera, Integrated photoelastic method-application to photoelastic isodynes, *Exp. Mech.*, *19*, 225–234 (1979).
36. R. K. Mueller and L. R. Saackel, Complete automatic analysis of photoelastic fringes, *Exp. Mech.*, 19, 245–251 (1979).
37. J. W. Dally, Dynamic photoelastic studies on fracture, *Exp. Mech.*, *19*, 349–361 (1979).
38. Y. Seguchi, Y. Tomita and M. Wanatabe, Computer-aided fringe-pattern analyser—A case of photoelastic fringe, *Exp. Mech.*, *19*, 362–370 (1979).
39. R. J. Sanford, Application of the least squares method to photoelastic analysis, *Exp. Mech.*, *20*, 192–197 (1980).
40. W. F. Swinson, J. L. Turner and W. F. Ranson, Designing with scattered light photoelasticity, *Exp. Mech.*, *20*, 397–402 (1980).
41. J. W. Dally, An introduction to dynamic photoelasticity, *Exp. Mech.*, *20*, 409–416 (1980).
42. J. Cernosek, Three dimensional photoelasticity by stress freezing, *Exp. Mech.*, *20*, 417–426 (1980).
43. R. J. Sanford, Photoelastic holography: a modern tool for stress analysis, *Exp. Mech.*, *20*, 427–436 (1980).
44. H. A. Gomide and C. P. Burger, Three-dimensional strain distribution in upset rings by photo-plastic simulation, *Exp. Mech.*, *21*, 361–370 (1981).
45. D. G. Berghaus, Simplification for scattered-light photoelasticity when using the unpolarised incident beam, *Exp. Mech.*, *22*, 394–400 (1982).
46. C. P. Burger and H. A. Gomide, Three-dimensional strain in rolled slabs by photo-plastic simulation, *Exp. Mech.*, *22*, 441–447 (1982).
47. R. Prabhakaran, Photo-orthotropic-elasticity: A new technique for stress analysis of composites, *Opt. Eng.*, *21*, 679–688 (1982).
48. C. W. Smith, W. H. Peters and G. C. Kirby, Crack-tip measurements in photoelastic models, *Exp. Mech.*, *22*, 448–453 (1982).
49. C. P. Burger and A. S. Voloshin, Half-fringe photoelasticity: a new instrument for whole field stress analysis, *ISA Trans*, *22*, 85–95 (1982).
50. R. B. Agarwal and L. W. Teufel, Epon 828 epoxy: a new photoelastic-model material, *Exp. Mech.*, *23*, 30–35 (1983).
51. K. A. Jacob, V. Dayal and B. Ranganayakamma, On stress analysis of anisotropic composites through transmission optical patterns: isochromatics and isopachics, *Exp. Mech.*, *23*, 49–54 (1983).

52. J. Komorowski and J. Stupnicki, Strain measurement with asymmetric oblique-incidence polariscope for birefringent coatings, *Exp. Mech.*, *23*, 171–176 (1983).
53. F. Zhang, S. M. Zhao and B. Chen, A digital image processing system for photoelastic stress analysis, *Proc. SPIE*, *814*, 806–809 (1987).
54. T. Y. Chen and C. E. Taylor, Computerised fringe analysis in photomechanics, *Exp. Mech.*, *29*, 323–329 (1989).
55. I. V. Zhavoronok, V. V. Nemchinov, S. A. Litvin, A. M. Skanavi, U. V. Pavlov and V. S. Evsenev, Automatization of measurement and processing of experimental data in photoelasticity, *Proc. SPIE*, *1554A*, 371–379 (1991).
56. J. T. Pindera, Scattered light optical isodynes basis for 3-D isodyne stress analysis, *Proc. SPIE*, *1554A*, 458–471 (1991).
57. A. Asundi and M. R. Sajan, Digital dynamic photoelasticity, *Opt. Lasers in Eng.*, *20*, 135–140 (1994).
58. M. Wolna, Polymer materials in practical uses of photoelasticity, *Opt. Eng.*, *34*, 3427–3432 (1995).
59. S. J. Haake, E. A. Patterson and Z. F. Wang, 2D and 3D separation of stresses using automated photoelasticity, *Exp. Mech.*, *36*, 269–276 (1996).
60. T. Kihara, A measurement method of scattered light photoelasticity using unpolarised light, *Exp. Mech.*, *37*, 39–44 (1997).
61. J.-C. Dupré and A. Lagarde, Photoelastic analysis of a three-dimensional specimen by optical slicing and digital image processing, *Exp. Mech.*, 37, 393–397 (1997).
62. G. Petrucci, Full field automatic evaluation of isoclinic parameter in white light, *Exp. Mech.*, 37, 420–442 (1997),

Moiré Phenomenon

1. F. K. Ligtenberg, The moiré method, a new experimental method for the determination of moments in small slab models, *Proc. Soc. Exp. Stress Anal.*, *12*, 83–98 (1954/55).
2. G. L. Rogers, A simple method of calculating moiré patterns, *Proc. Phys. Soc.*, *73*, 142–144 (1959).
3. D. L. A. Barber and M. P. Atkinson, Method of measuring displacement using optical gratings, *J. Sci. Instrum.*, *36*, 501–504 (1959).

4. J. M. Burch, The metrological applications of diffraction gratings, in *Progress in Optics*, *2*, pp. 73–108, (ed. Emil Wolf), North Holland, Amsterdam (1963).
5. G. Oster and Y. Nishijima, Moiré patterns, *Sci. Amer.*, *208*, 54–63 (1963).
6. P. S. Theocaris, Isopachic patterns by the moiré method, *Exp. Mech.*, *4*, 153–159 (1964).
7. G. Rieder and R. Ritter, Kruemmungsmessung an belasteter Platten nach dem Ligtenbergschen Moiré-Verfahren, *Forsch. Ing.-Wes.*, *31*, 33–44 (1965).
8. C. A. Sciammarella, Basic optical law in the interpretation of moiré patterns applied to the analysis of strains, Part 1, *Exp. Mech.*, *5*, 154–160 (1965).
9. B. E. Ross, C. A. Sciammarella and D. E. Sturgeon, Basic law in the interpretation of moiré patterns applied to the analysis of strains, Part 2, *Exp. Mech.*, *5*, 161–166 (1965).
10. D. Post, Analysis of moiré fringe multiplication phenomenon, *Appl. Opt.*, *6*, 1938–1942 (1967).
11. U. Heise, A moiré method for measuring plate curvature, *Exp. Mech.*, *7*, 47–48 (1967).
12. F. P. Chiang, Techniques of optical spatial filtering applied to processing of moiré fringe patterns, *Exp. Mech.*, *6*, 523–526 (1969).
13. H. H. M. Chau, Moiré pattern resulting from the superposition of two zone plates, *Appl. Opt. 8*, 1707–1712 (1969).
14. D. W. Meadows, W. O. Johnson and J. B. Allen, Generation of surface contours by moiré patterns, *Appl. Opt.*, *9*, 942–947 (1970).
15. H. Takasaki, Moiré topography, *Appl. Opt.*, *9*, 1467–1472 (1970).
16. J. der Hovanesian and Y. Y. Hung, Moiré contour-sum contour-difference and vibration analysis of arbitrary objects, *Appl. Opt.*, *10*, 2734–2738 (1971).
17. H. Takasaki, Moiré topography, *Appl. Opt.*, *12*, 845–850 (1973).
18. F. P. Chiang and G. Jaisingh, Dynamic moiré methods for the bending of plates, *Exp. Mech.*, *13*, 168–171 (1973).
19. K. Matsumoto and M. Takashima, Improvement on moiré technique for in-plane deformation measurements, *Appl. Opt.*, *12*, 858–864 (1973).
20. O. Bryngdahl, Moiré: formation and interpretation, *J. Opt. Soc. Amer.*, *64*, 1287–1294 (1974).

21. J. M. Burch and C. Forno, A high sensitivity moiré grid technique for studying deformation in large objects, *Opt. Eng.*, *14*, 178–185 (1975).
22. C. Chiang, Moiré topography, *Appl. Opt.*, 14, 177–179 (1975).
23. J. M. Walls and H. N. Southworth, The moiré pattern formed on superposing a zone plate with a grating or grid, *Opt. Acta*, *22*, 591–601 (1975).
24. J. Motycka, A grazing-incidence moiré interferometer for displacement and planeness measurement, *Exp. Mech.*, *15*, 279–281 (1975).
25. A. L. Browne and S. M. Rhode, Moiré contouring: Mathematical analysis, *Appl. Opt.*, *14*, 1260–1261 (1975).
26. F. P. Chiang, A shadow moiré method with two discrete sensitivities, *Exp. Mech.*, *15*, 382–385 (1975).
27. J. Marasco, Use of curved grating in shadow moiré, *Exp. Mech.*, *15*, 464–470 (1975).
28. Y. Yoshino and H. Takasaki, Doubling and visibility enhancement of moiré fringes of the summation type, *Appl. Opt.*, *15*, 1124–1126 (1976).
29. Y. Y. Hung, C. Y. Liang, J. D. Hovanesian and A. J. Durelli, Time-averaged shadow-moiré method for studying vibrations, *Appl. Opt.*, *16*, 1717–1719 (1977).
30. M. Idesawa, T. Yatagai and T. Soma, Scanning moiré method and automatic measurement of 3-D shapes, *Appl. Opt.*, *16*, 2152–2162 (1977).
31. A. Assa, A. A. Betser and J. Politch, Recording slope and curvature contours of flexed plates using a grating shearing interferometer, *Appl. Opt.*, *16*, 2504–2513 (1977).
32. C. Forno, Welds at high temperatures studied by moiré photography, *Weld and Met. Fabr.*, *46*, 661–667 (1978).
33. A. Assa, J. Politch and A. A. Betser, Slope and curvature measurement by a double-frequency-grating shearing interferometer, *Exp. Mech.*, *20*, 129–137 (1979).
34. D. T. Moore and B. E. Truax, Phase locked moiré fringe analysis for automated contouring of diffuse surfaces, *Appl. Opt.*, *18*, 91–97 (1979).
35. J. C. Perrin and A. Thomas, Electronic processing of moiré fringes: application to moiré topography and comparison with photogrammetry, *Appl. Opt.*, *18*, 563–574 (1979).

36. S. Yokozeki and S. Mihara, Moiré interferometry, *Appl. Opt.*, *18*, 1275–1280 (1979).
37. Y. Y. Hung, C. Y. Liang, J. D. Hovanesian and A. J. Durelli, Time averaged shadow moiré method for studying vibrations, *Appl. Opt.* *18*, 1717–1719 (1979).
38. R. E. Rowlands and P. O. Lemens, Moiré strain analysis in cryogenic environments, *Appl. Opt.*, 18, 1886–1887 (1979).
39. F. P. Chiang, Moiré methods of strain analysis, *Exp. Mech.*, *19*, 290–308 (1979).
40. M. L. Basehore and D. Post, Moiré method for in-plane and out-of-plane displacement measurement, *Exp. Mech.*, *21*, 321–328 (1981).
41. J. M. Burch and C. Forno, High resolution moiré photography, *Opt. Eng.*, *21*, 602–614 (1982).
42. C. A. Sciammarella, Holographic moiré an optical tool for the determination of displacements, strains, contours and slopes of surfaces, *Opt. Eng.*, *21*, 447–457 (1982).
43. L. Pirodda, Shadow and projection moiré techniques for absolute or relative mapping of surface shapes, *Opt. Eng.*, *21*, 640–649 (1982).
44. D. R. Andrews, Shadow moiré contouring of impact craters, *Opt. Eng.*, *21*, 650–654 (1982).
45. T. Y. Kao and F. P. Chiang, Family of grating techniques of slope and curvature measurements for static and dynamic flexure of plates, *Opt. Eng.*, *21*, 721–742 (1982).
46. S. Yokozeki, Moiré fringes, *Opt. Lasers in Eng.*, *3*, 15–27 (1982).
47. C. A. Sciammarella, The moiré method—a review, *Exp. Mech.*, *22*, 418–433 (1982).
48. T. Yatagai and M. Idesawa, Automatic fringe analysis for moiré topography, *Opt. Lasers in Eng.*, *3*, 73–83 (1982).
49. R. Ritter, Reflection moiré methods for plate bending studies, *Opt. Eng.*, *21*, 663–671 (1982).
50. J. Fujimoto, Determination of the vibrating phase by a time-averaged shadow moiré method, *Appl. Opt.*, *21*, 4373–4376 (1982).
51. T. Y. Kao and F. P. Chiang, Family of grating techniques of slope and curvature measurements for static and dynamic flexure of plates, *Opt. Eng.*, *21*, 721–742 (1982).
52. M. L. Basehore and D. Post, Displacement fields (U, W) obtained simultaneously by moiré interferometry, *Appl. Opt.*, *21*, 2558–2562 (1982).

53. R. Ritter and R. Schettler-Koehler, Curvature measurement by moiré effect, *Exp. Mech.*, *23*, 165–170 (1983).
54. G. Subramanian and S. Krishnakumar, On a curvature based nondestructive moiré technique for detecting defects in plates, *NDT Int. (UK)*, *16*, 271–274 (1983).
55. O. Kafri, A. Livnat and E. Keren, Optical second differentiation by shearing moiré deflectometry, *Appl. Opt.*, *22*, 650–652 (1983).
56. M. Halioua, R. S. Krishnamurthy, H. Liu and F. P. Chiang, Projection moiré with moving grating for automated 3-D topography, *Appl. Opt.*, *22*, 850–855 (1983).
57. L. Pirodda, Shadow and projection moiré techniques for absolute or relative mapping of surface shapes, *Opt. Eng.*, 640–649 (1983).
58. G. T. Reid, Moiré fringes in metrology, *Opt. Lasers in Eng.*, *5*, 63–93 (1984).
59. H. E. Cline, W. E. Lorensen and A. S. Holik, Automatic moiré contouring, *Appl. Opt.*, *23*, 1454–1459 (1984).
60. G. T. Reid, R. C. Rixon and H. I. Messer, Absolute and comparative measurements of three dimensional shape by phase measuring moiré topography, *Opt. Laser Technol.*, *16*, 315–319 (1984).
61. Y. Morimoto, T. Hayashi and N. Yamaguchi, Strain measurement by scanning—moiré method, *Bull. of JSME*, *27*, 2347–2352 (1984).
62. E. M. Weissman, D. Post and A. Asundi, Whole-field strain determination by moiré shearing interferometry, *J. Strain Anal.*, *19*, 77–80 (1984).
63. Y. Nakano and K. Murata, Measurement of phase objects using Talbot effect and moiré techniques, *Appl. Opt.*, *23*, 2296–2299 (1984).
64. L. Pirroda, Strain analysis by grating interferometry, *Opt. Lasers in Eng.*, *5*, 7 (1984).
65. J. Jahns, A. W. Lohmann and J. Ojeda-Castaneda, Talbot and Lau effects, a parageometrical approach, *Opt. Acta*, *31*, 313–324 (1984).
66. D. Post, R. Czarnek and D. Joh, Shear strain contours from moiré interferometry, *Exp. Mech.*, *25*, 282–287 (1985).
67. G. Subramanian and S. M. Nair, Direct determination of curvatures of bent plates using double-glass plate shearing interferometer, *Exp. Mech.*, *25*, 376–380 (1986).
68. K. Patorski and M. Kujawinska, Optical differentiation of displacement patterns using moiré interferometry, *Appl. Opt.*, *24*, 3041–3048 (1985).

69. A. Asundi, Variable sensitivity shadow moiré method, *J. Strain Anal.*, *20*, 59–61 (1985).
70. K. Patorski, Generation of derivatives of out-of-plane displacement using conjugate shear and moiré interferometry, *Appl. Opt.*, *25*, 3145–3151 (1986).
71. O. J. Lokberg and J. T. Malmo, Deformation measurements at very high temperatures by ESPI and moiré methods, *Proc. SPIE*, *661*, 62–68 (1986).
72. A. S. Voloshin, C. P. Burger, R. E. Rowlands and T. S. Richard, Fractional moiré strain analysis using digital image techniques, *Exp. Mech.*, 26, 254–258 (1986).
73. K. J. Gasvik and M. E. Fourney, Projection moiré using digital video processing. A technique for improving the accuracy and sensitivity, *Trans. ASME*, *108*, 652–656 (1986).
74. A. Asundi and M. T. Cheung, Moiré interferometry for out-of-plane displacement measurement, *J. Strain Anal.*, *21*, 51–54 (1986).
75. K. Andresen and D. Klassen, The phase-shift method applied to cross grating moiré measurements, *Opt. Lasers in Eng.*, *7*, 101–114 (1986/87).
76. K. Patorski, D. Post, R. Czarnek and Y. Guo, Real-time optical differentiation for moiré interferometry, *Appl. Opt.*, *26*, 1977–1982 (1987).
77. Y. Marimoto, Y. Seguchi and T. Higashi, Application of moiré analysis of strain using Fourier transform, *Proc. SPIE*, *814*, 295–302 (1987).
78. A. Asundi and M. T. Cheung, Three-dimensional measurement using moiré interferometry, *Strain*, *24*, 25–26 (1988).
79. M. Suzuki and M. Kanaya, Applications of moiré topography measurement methods in industry, *Opt. Lasers in Eng.*, *8*, 171–188 (1988).
80. C. Forno, Deformation measurement using high resolution moiré photography, *Opt. Lasers in Eng.*, *8*, 189–212 (1988).
81. Y. Morimoto, Y. Seguchi and T. Higashi, Application of moiré analysis of strain using Fourier Transform, *Opt. Eng.*, *27*, 650–656 (1988).
82. I. Glatt and O. Kafri, Moiré deflectometry-Ray tracing interferometry, *Opt. Lasers in Eng.*, *8*, 277–320 (1988).
83. J. M. Huntley and J. E. Field, High resolution moiré photography: application to dynamic stress analysis, *Opt. Eng.*, *28*, 926–933 (1989).

84. A. Asundi, Moiré interferometry for deformation measurement, *Opt. Lasers in Eng.*, *11*, 281–292 (1989).
85. Y. Guo, D. Post and R. Czarnek, The magic of carrier patterns, *Exp. Mech.*, *29*, 169–173 (1989).
86. D. Post, Determination of the thermal strains by moiré interferometry, *Exp. Mech.*, *29*, 318–323 (1989).
87. J. S. Lim, J. Kim and M. S. Chung. Additive type moiré with computer image processing, *Appl. Opt.*, *28*, 2677–2680 (1989).
88. M. E. Tuttle and D. L. Graesser, Compression creep of graphite/epoxy laminates monitored using moiré interferometry, *Opt. Lasers in Eng.*, *12*, 151–171 (1990).
89. Y. Sogabe, Y. Morimoto and S. Murata, Displacement measurement of track on magnetic tape by Fourier transform grid method, *Proc. SPIE*, *1554B*, 289–297 (1991).
90. C. G. Woychik and Y. Guo, Thermal strain measurements of solder joints in electronic packaging using moiré interferometry, *Proc. SPIE*, *1554B*, 461–470 (1991).
91. Y. Morimoto, D. Post and H. E. Gascoigne, Carrier pattern analysis of moiré interferometry using Fourier transform, *Proc. SPIE*, *1554B*, 493–502 (1991).
92. M. Kujawinska, L. Salbut and K. Patorski, 3-Channel phase stepped system for moiré interferometry, *Appl. Opt.*, *30*, 1633–1635 (1991).
93. C. G. Woychik and Y. Guo, Thermal strain measurements of solder joints in electronic packaging using moiré interferometry, *Proc. SPIE*, *1554B*, 461–470 (1991).
94. M. Suganuma and T. Yoshizawa, Three-dimensional shape analysis by the use of a projected grating image, *Opt. Eng*, *30*, 1529–1533 (1991).
95. Q. Yu, K. Andresen and D. S. Zhang, Digital pure shear-strain moiré patterns, *Appl. Opt.*, *31*, 1813–1817 (1992).
96. K. J. Gasvik and G. K. Robbersmyr, Autocorrelation of the setup for the projected moiré method, *Exp. Tech.*, *17*, 41–44 (1993).
97. J. Kato, I. Yamaguchi and S. Kuwashima, Real-time fringe analysis based on electronic moiré and its applications, in *Fringe '93*, (ed. W. Jüptner and W. Osten), pp. 66–71, Akademie Verlag (1993).
98. M. Priga, A. Kozlowska and M. Kujawinska, Generalization of the scaling problem for the automatic moiré and fringe projection

shape measurement systems, in *Fringe '93*, (eds. W. Jüptner and W. Osten), pp. 188–193, Akademie Verlag (1993).

99. J. T. Atkinson, and M. J. Lalor, Range measurement using moiré contouring, *Proc. SPIE*, *2340*, 202–210 (1994).
100. M. B. Whitworth and J. M. Huntley, Dynamic stress analysis by high-resolution reflection moiré, *Opt. Eng.*, *33*, 924–931 (1994).
101. P. K. Rastogi, Visualization of contours of equal slope of an arbitrarily shaped object using holographic moiré, *Opt. Eng.*, *33*, 2373–2377 (1994).
102. Y. Morimoto, H. E. Gascoigne and D. Post, Carrier pattern analysis of moiré interferometry using the Fourier transform method, *Opt. Eng.*, *33*, 2646–2653 (1994).
103. Y. Wang and F. P. Chiang, New moiré interferometry for measuring three-dimensional displacements, *Opt. Eng.*, *33*, 2654–2658 (1994).
104. A. K. Asundi, Chi-Shing Chan and M. R. Sajan, 360-deg profilometry: new techniques for display and acquisition, *Opt. Eng.*, *33*, 2760–2769 (1994).
105. S. K. Bhadra, S. K. Sarkar, R. N. Chakraborty and A. Basuray, Coherent moiré technique for obtaining slope and curvature of stress patterns, *Opt. Eng.*, *33*, 3359–3363 (1994).
106. T. Matsumoto, Y. Kitagawa and T. Minemoto, Sensitivity-variable moiré topography with a phase-shift method, *Opt. Eng.*, *35*, 1754–1760 (1996).
107. M. Wang, Projection moiré deflectometry for the automatic measurement of phase objects, *Opt. Eng.*, *35*, 2005–2011 (1996).
108. M. K. Kalms, W. Jueptner and W. Osten, Projected-fringe-technique with automatic pattern adoption using a programmable LCD-projector, in *Fringe '97*, (eds. W. Jüptner and W. Osten) pp. 231–236, Akademic Verlag (1997).

Index